Fifth International Conference on Building Materials and Construction (ICBMC 2020)

Innovations in Building Materials and Construction

IOP Conference Series: Materials Science and Engineering
Volume 829

Tokyo, Japan
26 - 29 February 2020

ISBN: 978-1-7138-2805-1
ISSN: 1757-8981

Printed from e-media with permission by:

Curran Associates, Inc.
57 Morehouse Lane
Red Hook, NY 12571

proceedings.com

Some format issues inherent in the e-media version may also appear in this print version.

Institute of Physics
Dirac House, Temple Back
Bristol BS1 6BE UK

Phone: 44 1 17 929 7481
Fax: 44 1 17 920 0979

techtracking@iop.org

Additional copies of this publication are available from:

Curran Associates, Inc.
57 Morehouse Lane
Red Hook, NY 12571 USA
Phone: 845-758-0400
Fax: 845-758-2633
Email: curran@proceedings.com
Web: www.proceedings.com

TABLE OF CONTENTS

SUSTAINABLE BUILDING

URBAN PLANNING

CONSTRUCTION ENGINEERING TECHNOLOGY

IOP Conf. Series: Materials Science and Engineering **829** (2020) 011001 doi:10.1088/1757-899X/829/1/011001

PREFACE

Welcome to the 5th International Conference on Building Materials and Construction, ICBMC 2020, held in Tokyo, Japan from February 26 to 29, 2020. ICBMC 2020 is at providing an avenue for academicians, students, researchers, professionals, engineers, and scientists from the academia and industry to share their research findings for further collaborative research in their respective areas.

The theme of the conference is "Innovations in Building Materials and Construction". proceedings of ICBMC 2020 consist of 27 selected papers from the conference which were submitted from universities, research institutes and industries. All papers were subjected to peer-review by conference committee members and international reviewers. The papers were selected based on their quality and relevancy to the conference.

I would like to express my gratitude and appreciation to the Conference Technical Committee for their great contribution in organizing this conference. Moreover, I wish to take this opportunity to express my heartfelt appreciation to plenary speakers, Prof. Nuno Dinis Cortiços from Universidade de Lisboa, Portugal, Prof. Kyoung Sun Moon Yale University, USA, and Prof. Ho Chee Cheong, Universiti Tunku Abdul Rahman, Malaysia, for sharing their knowledge with the rest of the participants. I hope the experience of ICBMC 2020 was intellectually stimulating to all participants.

Last, I hope that the contents of this volume will prove useful for researchers and practitioners in developing and applying new technologies and processes in the field of building materials and construction.

Prof. Tan Kiang Hwee
National University of Singapore, Singapore
March 11, 2020

IOP Conf. Series: Materials Science and Engineering **829** (2020) 011002 doi:10.1088/1757-899X/829/1/011002

Conference Committee

Advisory Chair

Prof. Tan Kiang Hwee, National University of Singapore, Singapore

Conference Chair

Prof. Umemura Kazuo, Tokyo University of Science, Japan

Program Chairs

Prof. Nuno Dinis Cortiços, University of Lisbon, Portugal

Assoc. Prof. Kyoung Sun Moon, Yale University, USA

Steering Chair

Prof. Ho Chee Cheong, Universiti Tunku Abdul Rahman, Malaysia

Technical Committee

Prof. Mahen Mahendran, Queensland University, Australia

Prof. Zbyšek Pavlík, Czech Technical University, Czech Republic

Prof. Ippei Maruyama, Nagoya University & The University of Tokyo, Japan

Prof. Sang Whan Han, Hanyang University, South Korea

Prof. Ludovic JASON, SEMT, CEA DEN, Université Paris Saclay, France

Prof. Yibo Liu, Beijing Gang Yan Diamond Products Company, China

Prof. Muhammad Harunur Rashid, Khulna University of Engineering & Technology, Bangladesh

Assoc. Prof Galina Gorbacheva, Mytishchi Branch of Bauman Moscow State Technical University, Russia

Assoc. Prof Dmitry Levushkin, Mytishchi Branch of Bauman Moscow State Technical University, Russia

Assoc. Prof Liang Xu, Beijing Gang Yan Diamond Products Company, China

Assoc. Prof. Ye Guang, Delft University of Technology, Netherlands

Assoc. Prof. Toshiaki Sato, Kyushu University, Japan

Assoc. Prof. Zihni TURKAN, Near East University, Cyprus

Assoc. Prof. Hans Beushausen, University of Cape Town, Cape Town, South Africa

Assoc. Prof. Amnon Katz, Technion, Israel

Assoc. Prof. Dr. Mohammed N. Abdulrazaq Alshekhly, Management & Science University, Malaysia

Assoc. Prof. Mohammad Arif Kamal, Aligarh Muslim University, India

Assoc. Prof. Maatouk Khoukhi, United Arab Emirates University, UAE

Dr. Fernando Pachego Torgal, University of Minho, Portugal

Dr. Doo-Yeol Yoo, Hanyang University, South Korea

Dr. Trong-Phuoc Huynh, Can Tho University, Vietnam

Dr. B. Kondraivendhan, S. V., National Institute of Technology, India

Dr. Baydaa Hussain Maula, Middle Technical University, Iraq

Dr Ahmed Ajel Ali, Kufa university, Iraq

Dr Kedsarin Pimraksa, Chiang Mai University, Thailand

Dr. Wegdan Wagdy El-Nadoury, Pharos University, Egypt

IOP Conf. Series: Materials Science and Engineering **829** (2020) 011003 doi:10.1088/1757-899X/829/1/011003

Peer review statement

All papers published in this volume of *IOP Conference Series: Materials Science and Engineering* have been peer reviewed through processes administered by the proceedings Editors. Reviews were conducted by expert referees to the professional and scientific standards expected of a proceedings journal published by IOP Publishing.

IOP Conf. Series: Materials Science and Engineering **829** (2020) 012001 doi:10.1088/1757-899X/829/1/012001

Sustainable Concrete in Transportation Infrastructure: Australian Case Studies

Koorosh Gharehbaghi[1,a], Farshid Rahmani[1,b] and David Paterno[1,c]

[1]RMIT University, Melbourne, Australia

[a] koorosh.gharehbaghi@rmit.edu.au, [b] farshid.rahmani@rmit.edu.au, [c] david.paterno@rmit.edu.au

Abstract. Whilst generally concrete and steel are the most common substances for civil engineering construction, new trends and development such as sustainable and green material have a lesser effect on the environment. Typically, approximately 90% of a concrete structure's CO_2 emissions are a result of the energy consumed during its life, there is much that can be done to reduce that 10% associated with its construction. Thus, innovative engineering materials, such as Recycled Concrete (RC) have emerged with the potential to influence the future of an environmentally sustainable construction industry. This research investigates the RC as the basis of sustainable concrete for civil engineering construction such as roads, rail and so on. Furthermore, number of examples, 6 roads and 6 rail segments in Sydney, were reviewed. Overall, this research found that for the 6 road and rail case studies generally only RC is utilized as the basis of sustainable material. Although, this is a very small sample, however, a pattern can be noticed. Unfortunately, the pattern is the lack of innovative sustainable material other than concrete for such transportation infrastructure. Appropriately, further investigation is required to review the possibility of other sustainable materials including those incorporating waste-by products as the basis consolidated sustainable material usage in transportation infrastructure.

1. Introduction

The general aim of sustainable materials is to expend less energy, and to reduce environmental impacts [1-4]. Utilizing such materials is detrimental to ensure a sustainable living. An example of such material is Recycled Concrete (RC). Typically, approximately 90% of a concrete structure's CO_2 emissions are a result of the energy consumed during its life, there is much that can be done to reduce this [5-7]. Gharehbaghi [8] noted that, within the building and construction field, concrete is the primary starting point for the CO_2 emissions reduction. This is because concrete is responsible for approximately 10% of global CO_2 emission [9-11]. For sustainable concrete, more greener forms of substance can be formed which utilizes recycled materials in its mix [12,13]. For example, crushed glass can be added, as can wood chips or slag – a byproduct of steel manufacturing. Whilst these changes do not fundamentally transform concrete, however, using such processing methods would ultimately reduce the CO_2 emissions associated such material manufacturing. As shown in Fig 1, the use of concrete is on-going regardless of its CO_2 emissions footprint.

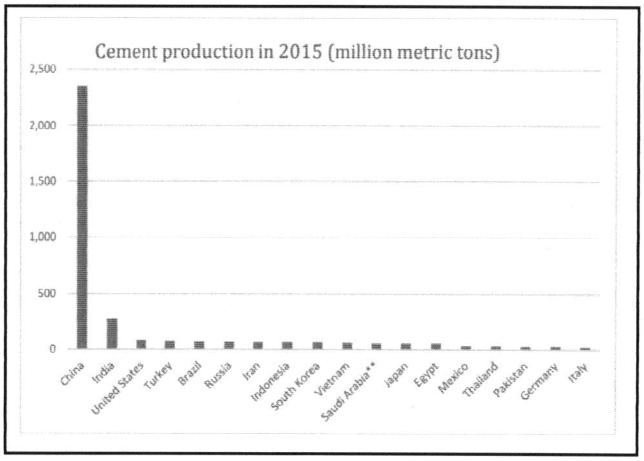

Figure 1. concrete use by individual nations [38].

For emerging economies including China and India, concrete is still an essential material. Such popularity is due to concrete's performances for different conditions and climates together with its ease of use [14-16]. Nonetheless, the concrete's environmental impact is still an on-going concern. Thus, innovative engineering materials, such as RC have emerged with the potential to influence the future of an environmentally sustainable construction product industry. Further, for the transportation infrastructure such modern material processing will be also significantly beneficial.

2. Literature review

2.1. Sustainable concrete in transportation infrastructure

Although there are various systems available to utilize in material selection, processes are not universal, and the importance of factors alters depending on the country of location [17-20]. Fehling et al. [21] identified multiple barriers to incorporate sustainability into material selection practices. These include perceived costs, time to source materials, education and training, understanding in-house experts. Akadiri [22] further investigated the barriers of implementing a sustainable selection practice and mentioned that a lack of sustainable material information, perception of extra costs being incurred, perception of extra time being incurred, perception that sustainable materials are low in quality and an unwillingness to change the conventional way of specifying are the most influential barriers affecting the incorporation of sustainable design within high-rise construction. A review of literature revealed cost figured predominantly as a barrier to achieving sustainability [2,23,24]. In selecting sustainable concrete, benefits such as the overall productivity, performance and so on need to be considered [25,3]. Chevalier et al. [26] argued that specific factors to consider when using sustainable concrete such as RC, include its functioning capabilities among others. The sustainable concrete's functioning ability aligns with its structural capabilities [27]. This includes load bearing capability, stress and strain probabilities and so on. Gharehbaghi and Georgy [28] noted that, the greatest problem facing sustainable concrete is early deformation and degradation. Transportation infrastructure and their subsequent connection systems use concrete in their assembly and configuration including structural support and so on [29]. While for roads, concrete is extensively utilized for decks and carriageway surfacing, for railways it is used for sleepers and other structural systems. Subsequently, recycled concrete aggregates would be sustainably favorable for transportation infrastructure [30]. However, one of the greatest concerns of recycled concrete is its durability and performance. Such concern also includes using recycled concrete for transportation infrastructure [31,32].

Mamlouk and Zaniewski [33] along with Sadeket al. [23] and, Singh et al. [17] noted that, some of the concerns in using recycled concrete, significant deterioration rate due to excessive exposure to substantial water, significant temperature changes and air contaminants which all are common in transportation infrastructure. Furthermore, table 1 summarizes some risk associated with sustainable concrete in transportation infrastructure. However, based on the examination of the literature, it is therefore evident that engineers do not perceive that there is a widespread demand for sustainable concrete in transportation infrastructure. Accordingly, a review into new trends and developments for such infrastructure needs to be undertaken.

2.2. New trends and developments

Hibbeler and Vijay [34] together with Öchsnerand Altenbach [35] as well as Jahan et al. [36] highlighted that traditional material selection for transportation infrastructure generally include longevity. Nonetheless, over the past few years many recycled materials including RC are been used in such infrastructure [37]. Specifically, recycled cement, plastic and asphalt are being used more and more [39]. These sustainable materials are used for low use areas such as local roads [39, 40]. However, for major arterial road together with rail networks currently, only RC seems to be the best option [41]. The reason for such limitation of material selection is based on: highly complex design of the structure together with exceedingly cost due to the size of roads and rail networks [42,43]. For these reasons the use of sustainable concrete in transportation infrastructure is still unfortunately limited. Dejectedly this has led to lack of available sustainable materials for such infrastructure. Due to high performance concrete composition together with long-term material performance, the RC is still the only major sustainable concrete in transportation infrastructure. To support such claim 6 road and six rail segments have been studied.

3. Research methodology

The research will use qualitative method and therefore utilises Australian case studies to review new trends of sustainable concrete in transportation infrastructure. In achieve such outcome, document search was undertaken to define the sustainable concrete specifics of road and rail infrastructure.

4. Case studies

As the basis of case studies, 6 roads and 6 rail segments in Sydney, are investigated. Both tables 1 and table 2 present the overview of these case studies.

Table 1. Overview of roads case studies.

Road segment #	Road classification	Sustainable material and portion	Sections used
SydEast0097	Major collector road	Recycled concrete <10%	Bituminous layer
SydNorthEast0024	Major collector road	Recycled concrete <10%	Bituminous layer
SydSouthEast013	Principal arterial	Recycled concrete <3%	Bituminous layer
SydWest0016	Minor arterial	Recycled concrete <5%	Bituminous layer
SydSouthEast008	Minor arterial	Recycled concrete <5%	Bituminous layer
SydNorthEast0151	Local road	Recycled concrete <15% and plastic <5%	Bituminous and surface layers

As observed, for the 6 road segments only a small portion possess sustainable materials. For such infrastructure recycled concrete were the main composition of sustainable materials. While the Bituminous base course consisted of traditional Sub-grade gravel, stones and sands; the Binder course also possessed conventional materials. On the other hand, Bituminous concrete layer is where all the sustainable material was utilised. For theses six case studies, recycled concrete was the only material used. Only for SydNorthEast0081 recycle plastic was used as basis of surface covering.

Table 2. Overview of rails case studies.

Rail segment #	Rail classification	Sustainable material and portion	Sections used
T1032	Class 3: Local railroad	Recycled concrete <10%	Rail sleepers
T4016	Class 3: Local railroad	Recycled concrete <10%	Rail sleepers
T5009	Class 3: Local railroad	Recycled concrete <10%	Rail sleepers
T5019	Class 3: Local railroad	Recycled concrete <10%	Rail sleepers
T7005	Class 2: Switching terminal railroad	Recycled concrete <10%	Rail sleepers
T9037	Class 3: Local railroad	Recycled concrete <10%	Rail sleepers

Sydney currently has 9 main rail network lines, ranging from T1 to T9. For these, standard Australian intermediate gauge railways are used. In addition, for these 6 case studies, steel is also used as the basis of rails, while recycled concrete is utilized for sleepers along with stones for ballast. Overall, as it can be noticed, for these 6 road and rail case studies generally only RC is utilized as the basis of sustainable material. Although, this is a very small sample, however, a pattern can be noticed. Unfortunately, the pattern is the lack of innovative sustainable material other than concrete for such transportation infrastructure.

5. Conclusion and recommendations

Concrete together with steel form part of the primary materials used in transportation infrastructure. Due to ever increase in CO_2 emissions, new sustainable materials are sought for such infrastructure. Innovative engineering materials, including RC are the back-bone of greener propositions for such environmental considerations. Subsequently, this research found that the new trends and developments for transportation infrastructure is limited mainly to RC. To demonstrate such dilemma, 6 Australian road and six rail segments were reviewed. Due to such limitation new and improved substance other than RC are essential to embrace sustainable material specifically for transportation infrastructure. Areas for further research could include the investigation of other sustainable materials including those incorporating waste-by products. Other sustainable materials which are in testing phase or have recently been developed, could further promote sustainability in transportation infrastructure.

6. References

[1] R. Riles, M. Bilec, N. Gokhan, and K. Needy: The Economic Benefits of Green Buildings: A Comprehensive Case Study, The Engineering Economist, Volume 51. Issue 3 (2006), pp 259-295.

[2] Z. Bribian, V. Capilla, and A. Uson: Life cycle assessment of building materials: Comparative analysis of energy and environmental impacts and evaluation of the eco-efficiency improvement potential', Building and Environment, vol. 46 (2011), pp. 1133-1140.

[3] K. Gharehbaghi, and F. Rahmani: Practicalities and Developments of High-Rise Composite Structures: Case Studies, Materials and Technologies in Engineering (2018), pp 153-159.

[4] K. Reddy, C. Cameselle, and J. Adams: Engineering Practices, Sustainable Engineering: Drivers, Metrics, Tools, Engineering Practices, and Applications, Vol. 1, no.1 (2019), pp 419.

[5] J. Glasson, R. Therivel, and A. Chadwick: Introduction to Environmental Impact Assessment, Taylor and Francis, UK (2012), pp 87-96.

[6] M. Gere, and J. Goodno: Mechanics of materials, 8th edition, Stamford: Cengage Learning. (2013).

[7] A. Lukman, O. AkanbiacLukumon, O. Olugbenga, A. Akinadea, A. AjayiaJuan, M. Davila, D. Hakeem, and A. Owolabia: Disassembly and deconstruction analytics system (D-DAS) for construction in a circular economy, Journal of Cleaner Production, Vol. 223. (2019) pp. 386-396.

[8] K. Gharehbaghi: Advancements in concrete technology in Australia: geo-polymer concrete, International Journal of Constructed Environment, volume 7, issue 1 (2015), pp 19-29.

[9] A. Arıoğlu, Ö. Mahmure, D. Dhavale, and , J. Sarkis: Greenhouse gas emissions in the construction industry: An analysis and evaluation of a concrete supply chain, Journal of Cleaner Production, vol.167 (2017), pp.1195-1207.

[10] K. Gharehbaghi, and C. Scott-Young: Waste diminution in construction projects: Environmental predicaments, Earth and Environmental Science, Institute of Physics, United Kingdom, vol. 127, no. 1. (2018b) pp. 58-63.

[11] M. Eckelman, C. Brown, L. Troup, L. Wang, M. Webster, and J. Hajjar: Life cycle energy and environmental benefits of novel design-for-deconstruction structural systems in steel buildings ,Building and Environment, Vol.143 (2018), pp.421.

[12] J. Giesekam, J. Barrett, and P. Talyor: Construction sector views on low carbon building materials, Building Research & Information, vol 44 (2015) pp. 423-444.

[13] K. Gharehbaghi, and R. Chenery: Fiber Reinforced Concrete (FRC) for High Rise Construction: Case Studies, Materials Science and Engineering, vol. 272 (2017),pp 223-229.

[14] E. Kocs: The Global Carbon Nation: Status of CO2 Capture, Storage and Utilization, EPJ Web of Conferences 2017, Les Ulis: EDP Sciences, Vol 148 (2017) pp. 3-31.

[15] S. Lockrey, K. Verghese, E. Crossin, and H. Nguyen: Concrete recycling life cycle flows and performance from construction and demolition waste in Hanoi, Journal of Cleaner Production, vol. 179 (2018), pp. 593-604.

[16] V. Kashintseva, W. Strielkowski, J. Streimikis, and T. Veynbender: Consumer attitudes towards industrial CO_2 capture and storage products and technologies, Energies, vol. 11, no. 10. (2018).

[17] M. Singh, T. Ohji, and R. Asthana: Progress and Prospects, Green and Sustainable Manucturing of Advanced Materials, 1stedn, (2015), pp. 3-10, Elsevier.

[18] C. Kilbert: Sustainable construction: Green building design and delivery, Hoboken, Wiley, (2016), pp 78-82.

[19] N. Bestner: Concrete for high-rise buildings: Performance requirements, mix design and construction considerations, Structural Concrete Properties and Practice, (2013), pp. 1-4.

[20] K. Gharehbaghi, and C. Scott-Young: GIS as a vital tool for Environmental Impact Assessment and Mitigation, Earth and Environmental Science, Institute of Physics, United Kingdom, vol. 127, no. 1, (2018a), pp. 52-57.

[21] E. Fehling, B. Middendorf, and J. Thiemicke: Ultra-High Performance Concrete and High Performance Construction Materials, Kassel University Press, Kassel, Germany (2016).

[22] O. Akadiri: Understanding barriers affecting the selection of sustainable materials in building projects, Journal of Building Engineering, vol. 4 (2015), 86-93, Elsevier.

[23] M. Sadek, M. El-Attar, and A. Ali: Reusing of marble and granite powders in self-compacting concrete for sustainable development, Journal of Cleaner Production, vol. 121, (2015), pp., 19-32, Elsevier

[24] B. Guerra, A. Bakchan, F. Leite, and K. Faust: BIM-based automated construction waste estimation algorithms: The case of concrete and drywall waste streams, Waste Management, Vol. 87 (2019), pp.825-832.

[25] A. Yates: Sustainable Infrastructure – Sustainable Buildings, ICE Publishing (2015), pp 121-134.

[26] Y. Chevalier, and J. Vinh Tuong: Mechanics of viscoelastic materials and wave dispersion, London : Hoboken: Wiley (2010).

[27] W. Findley, and F. Davis: Creep and Relaxation of Nonlinear viscoelastic materials, Newburyport : Dover Publications (2013).

[28] K. Gharehbaghi, and M. Georgy: Sustainable Construction by Means of Improved Material Selection Process, International journal on the Academic Research Community publication, vol 1, issue 1 (2019), pp 85-94.

[29] H. Zhou, S. Sung, G. Li, and Y. Zie: Topology optimization for microstructures of viscoelastic composite materials, Elsevier (2015).

[30] P. Xincheng: Super-high-strength high performance concrete, Hoboken: Taylor and Francis (2012).

[31] L. Ghiringhelli, M. Terraneo, and E. Vigoni: Improvement of structures vibroacoustics by widespread embodiment of viscoelastic materials, Aerospace Science and Technology, Vol.28(1), (2013), pp.227-241

[32] A. MuntasirBillah, and M. ShahriaAlam: Seismic fragility assessment of highway bridges: a state-of-the-art review, Structure and infrastructure engineering, volume 11, issue 6 (2015).

[33] M. Mamlouk, J. Zaniewski: Materials for civil and construction engineers, Third edition, Prentice Hall (2011).

[34] C. Hibbeler, and S. Vijay: Mechanics of materials, 9th edition, worldwide adaptation edition, SI edition, Singapore: Pearson Education South Asia Pty Ltd (2014).

[35] A. Öchsner, and H. Altenbach: Mechanical and Materials Engineering of Modern Structure and Component Design, Springer International Publishing (2015).

[36] A. Jahan, L. Edwards, and M. Bahraminasab: The importance of decision support in materials selection, Multi-criteria Decision Analysis for Supporting the Selection of Engineering Materials in Product Design (2016), pp. 1-23, Elsevier.

[37] W. Salama: Design of concrete buildings for disassembly: An explorative review, International Journal of Sustainable Built Environment, vol. 6, no. 2 (2017), pp. 617-635.

[38] L. Xing: The Rise and Potential Peak of Cement Demand in the Urbanized World Report, 'CORNERSTONE, The official Journal of world's coal industry (2015).

[39] Gharehbaghi, K., McManus K., and Robson, K., (2019), "Minimizing the environmental impacts of mega infrastructure projects: Australian public transport perspective", Journal of Engineering, Design and Technology, vol. 17, issue 4, pp. 736-746.

[40] Domone, P., Illston, J., (2010), "Construction materials: their nature and behaviour", 4th edition, Spon Press.

[41] Gharehbaghi, K., Georgy, M., and Rahmani, F., (2018), "Composite High-Rise Structures: Structural Health Monitoring (SHM) and Case Studies", Materials and Technologies in Engineering, pp 146-152.

[42] Myers, J., (2007), 'The Use of High Strength/High Performance Concrete in America: A Code and Application Perspective', Department of Civil, Architectural and Environmental Engineering, vol. 1, pp. 1-10.

[43] Skrzypek, Jacek and Ganczarski, Artur: Mechanics of Anisotropic materials, Springer International Publishing (2015).

Performance of FRC and GPC for High-Rise Construction: Case Studies

Koorosh Gharehbaghi[1,a], Farshid Rahmani[1,b], David Paterno[1,c] and Sara Gharehbaghi[2,d]

[1]RMIT University, Melbourne, Australia
[2]Bechtel Corporation, San Francisco, United States

[a] koorosh.gharehbaghi@rmit.edu.au, [b] farshid.rahmani@rmit.edu.au, [c] david.paterno@rmit.edu.au, [d]sara.eng.gharehbaghi@gmail.com

Abstract. This paper aims to compare Fiber Reinforced Concrete (FRC) and Geo-Polymer Concrete (GPC) for high-rise construction. In doing so, 8 case studies from the USA and Australia are reviewed. The ever-increasing use of material processing linked to environmental degradation has encouraged industries to seek alternative materials. Both FRC and GPC are examples of such alternative materials. Their composition of these materials is derived from manufacturing, building and construction excess. Concerns remain, however, regarding these material's abilities such as strength, when compared to the traditional i.e. concrete, steel etc. This paper reviews select case studies where FRC and GPC materials have been used. In addition, their overall structural integrity was also observed. This paper then scrutinized their Structural Health Monitoring (SHM) to uncover their specific structural performance. Subsequently, this paper found that for both FRC and GPC extreme weathering conditions may affect their structural performance. Finally, case study review also indicates additional comprehensive research is required to review such relationship.

1. Introduction

Gharehbaghi and Scott-Young [1] argued that due to on-going environmental degradation, comprehensive environmental impact assessment and mitigation planning is desirable for all construction projects. Environmental degradation has indeed proved bothersome for engineers and planners [2-5]. Indeed, such dilemmas are one main driver for seeking alternative construction materials [6-9]. FRC and GPC are two leading examples of alternative materials.

FRC is a material which utilizes Fiber as its bonding agents [10]. GPC on the other hand, makes use of polymers to replace cement as a binder in concrete [5,11]. Cuerca [12] highlighted that both of these innovative materials are increasingly utilized in high rise construction. Researchers including Gustavo [13] and Hugo [14] demonstrated that both materials can withstand required structural loads. Conversely, such both FRC and GPC are also subject to some skepticism when compared with the traditional Reinforced Concrete (RC) [15]. In light of these competing views, this research aims to provide additional microscopic comparison of both FRC and GPC for the high-rise construction. In doing so, 8 case studies from USA and Australia will be examined.

IOP Conf. Series: Materials Science and Engineering **829** (2020) 012002 doi:10.1088/1757-899X/829/1/012002

2. Literature review and background

Gharehbaghi and Cheany [10] propositioned that establishing structural performance is a vital issue to guide the use of FRC. Although there are many literatures discussing such focus, the most important aspect of FRC is structural performance [16-20]. Questions around the wide-scale use of FRC stem from FRC mechanical composition [21,22]. In addition, Gharehbaghi et al. [23] also highlighted that FRC bonding processes such as fiber type, i.e. steel, glass are important aspects of structural performance.
On the other hand, unlike Portland Cement (PC), GPC were initially used in producing precast and pre-stressed concrete [24]. Gharehbaghi [5] , highlighted that, initially GPC involved developing ultra-quick early high strength cement with room temperature curing. He further propositioned that, such procedure has the ability to allow for speedier bonding process. In addition, while the primary advantage of GPC is greenhouse gas emission reductions, it also provides the potential to utilize waste materials and processing by-products. Nonetheless, GPC possess a good strength and durability characteristics. Obstacles for GPC usage remain, however, and include, industry unenthusiastic perceptions which lead to statutory and regulatory issues [15,25,26].
Due to ever increase of environmental impacts because of manufacturing processes together with increase cost in such industrialization, alternative materials such as FRC and GPC are gradually developing into alternative solutions [27-31]. Structural Health Monitoring (SHM) is one method to simplify the use of FRC and GPC for high-rise construction. SHM generally refers to damage detection and preservation regimes. It is determined to act as a good structural monitoring technique (Hibbeler and Sekar, 2014). SHM can also be used to determine FRC and GPC performance for high-rise construction. SHM is comprehensive and can also determine the overall wear-and-tear rate of structures [32,33]. Succar et al. [34] noted that high-rise structures in particular are subject to long-term life cycles - their main cause of deterioration. Gharehbaghi and Scott-Young [35] suggested that SHM might even be detrimental for unconventional materials. Since both FRC and GPC can be categorized as such materials, SHM is presently problematic. Inventive damage detection strategies together with modern and simulated SHM would thus beneficial for FRC and GPC [37-39]. Importantly, both FRC and GPC may be subject to additional degradation processes including excessive heat, weathering and so on. Nevertheless, for high-rise construction, both FRC and GPC can ultimately provide an economical option where compared to RC and PC. However, a careful investigation needs to be carried out to further compare FRC and GPC.

3. Research methodology

The research conducted follows both qualitative and quantitative approaches. The main methodology employed, however, is a case studies approach. Taking high-rise construction as their focus, these case studies are located in the USA and Australia. Other methodology that will be used to further enhance the research include archival analyses. Subsequently, most of the data will be archival analysis and data examination. This include, log-books, SHM schematics and reports and so on. In addition, industry-based reports were also accessed to further appraise these findings.

4. Case studies

To systematically compare FRC and GPC for high rise construction, 8 case studies of high-rise buildings were evaluated. A summary of the case studies is provided in table 1.

Table 1. Comparison of case studies.

Structure	Age (years)	Number of storeys	FRC	GP	SHM performance
USA-Buil[1]	5	7	Floor slabs	External panels	Minor surface cracks on one of the floors
USA-Buil[2]	6	8	External walls	Internal walls	Satisfactory and enduring

USA-Buil[3]	3	5	Limited panels	Internal walls	Ideal
USA-Buil[4]	4	6	Internal walls	Floor slabs	Minor structural degradation, maintenance regime commenced
AUST-Buil[1]	4	7	External walls	Internal walls	Satisfactory and enduring
AUST-Buil[2]	5	8	Internal walls	Internal walls	Satisfactory and enduring
AUST-Buil[3]	3	8	External walls	Internal walls	Satisfactory and enduring
AUST-Buil[4]	4	8	Internal walls	Floor slabs	Ideal

Overall, both FRC and GPC were utilized for either non-load or low-load bearing structural components. The SHM were further utilized to recommend maintenance regimes and inspection intervals. Sensor Connect Software was also used for specific structural integrity analysis. Most of the buildings reviewed are new- <7 years old. Furthermore, building SHM performance were predominately and consistently satisfactory. SHM logs were used to collect sensor data from all 8 buildings. Such data was then analyzed on a high-end platform. All the buildings utilized Steel as the source of their FRC. Importantly, all the buildings were Monocoque structures. USA-Buil4 though, had some minor structural performance issues, due to extreme weather alteration between cold to hot.
As Gharehbaghi and Rahmani [36] noted, both FRC and GPC may demonstrate similar inadequacies when exposed to continuous, extreme altering weathering. Although one such case found in this research, this is not a strong vindication of such proposition. Moreover, the fiber bonding process could also be affected by extreme altering temperatures. On the other hand, the rest of the buildings were all satisfactory and enduring, although USA-Buil1 had some minor surface cracks. To further assess these case studies, their geographical locality and environmental conditions need to be also investigated.

5. Findings

Figure 1 shows the result of the weathering condition investigation.

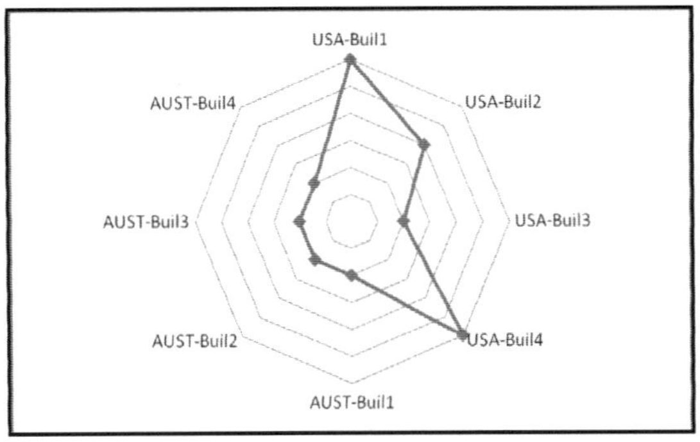

Figure 1. Weathering condition comparison.

Whereas mild weather alterations include predictable seasonal conditions, severe variations include extremes of the historical distribution such as stern temperature discrepancy and so on. Further, the geographical locality and building location also influenced the weathering conditions. As it can be noticed both USA-Buil1 and USA-Buil4 were subject to severe weather variations, particularly temperature changes. These extreme temperature changes were estimated between (approximately) -15co to +40co. Such weather changes are abnormal for the specific geographical locality, and therefore might explain the SHM result for both USA-Buil1 and USA-Buil4. Nonetheless, USA-Buil1 seems to

IOP Conf. Series: Materials Science and Engineering **829** (2020) 012002 doi:10.1088/1757-899X/829/1/012002

outperform USA-Buil4.Close investigation found that USA-Buil1 had additional structural reinforcement especially for Seismic activities. Finally, Sensor Connect analysis results were used to recommend appropriate maintenance and rehabilitation regimes to prolong the buildings life.

6. Conclusion

Although not new, both FRC and GPC are gradually gaining popularity in building industry predominantly for high rise construction. Although there have been numerous studies in justifying the Fiber Reinforced Concrete (FRC) for general construction, this paper provided several case studies (from Australian and United States) to support its usage specifically for high rise construction. This paper also reviewed the performance of FRC and GPC specifically for high-rise construction. Concerns, however, regarding the material's strength abilities when compared to the traditional i.e. concrete, steel etc arose. Although, both FRC and GPC have been researched extensively, this paper used case studies further review the performance of both materials. Overview, SHM analysis discovered, in 8 buildings sampled, that extreme weather conditions may impact the performance of both FRC and GPC structures. The review present above indicates the need for, more comprehensive research which includes bigger sample sizes to review such relationship.

7. References

[1] K. Gharehbaghi, and C. Scott-Young: GIS as a vital tool for Environmental Impact Assessment and Mitigation, Earth and Environmental Science, Institute of Physics, United Kingdom, vol. 127, no. 1, (2018a), pp. 52-57.

[2] J. Glasson, R. Therivel, and A. Chadwick: Introduction to Environmental Impact Assessment, Taylor and Francis, UK (2012), pp 87-96.

[3] N. Bestner: Concrete for high-rise buildings: Performance requirements, mix design and construction considerations, Structural Concrete Properties and Practice, (2013), pp. 1-4.

[4] A. MuntasirBillah, and M. ShahriaAlam: Seismic fragility assessment of highway bridges: a state-of-the-art review, Structure and infrastructure engineering, volume 11, issue 6 (2015).

[5] K. Gharehbaghi: Advancements in concrete technology in Australia: geo-polymer concrete, International Journal of Constructed Environment, Volume 7, issue 1 (2015), pp 19-29.

[6] L. Courard: Recycled and bio-based materials for sustainable construction, in Symposium 3R: Reduce, Reuse, Recycle in Constructions Conference, Warsaw, Poland (2016).

[7] D. Densley, and T. Davison: Developing an LCA methodology to account for the environmental benefits of design for deconstruction, Building and Environment, Vol.57 (2012), pp. 387-395.

[8] P. Lieblang: Sustainable Construction with Concrete, Modern Concrete Construction Manual, Structural Design, Material Properties, Sustainability, Vol. 1, no. 1 (2014), pp. 116-129.

[9] K. Gharehbaghi, and M. Georgy: Sustainable Construction by Means of Improved Material Selection Process, International journal on the Academic Research Community publication, vol 1, issue 1 (2019), pp 85-94.

[10] K. Gharehbaghi, and R. Chenery: Fiber Reinforced Concrete (FRC) for High Rise Construction: Case Studies, Materials Science and Engineering, vol. 272 (2017),pp 223-229.

[11] E. Fehling, B. Middendorf, and J. Thiemicke: Ultra-High Performance Concrete and High Performance Construction Materials, Kassel University Press, Kassel, Germany (2016).

[12] E. Cuerca: On Shear Behavior of Structural Elements made of Steel Fiber Reinforced Concrete, SpringerLink (2014).

[13] J. Gustavo: High Performance Fiber Reinforced Cement Composites, 6th International RILEM conference on high performance fiber reinforced cement composition (2012).

[14] H. Hugo: Performance Based Building Design 1 From Below Grade construction to Cavity Walls, Hoboken: Wiley (2012).

[15] K. Reddy, C. Cameselle, and J. Adams: Engineering Practices, Sustainable Engineering: Drivers, Metrics, Tools, Engineering Practices, and Applications, Vol. 1, no.1 (2019), pp 419.

[16] R. Riles, M. Bilec, N. Gokhan, and K. Needy: The Economic Benefits of Green Buildings: A Comprehensive Case Study, The Engineering Economist, Volume 51. Issue 3 (2006), pp 259-295.
[17] S. Lockrey, K. Verghese, E. Crossin, and H. Nguyen: Concrete recycling life cycle flows and performance from construction and demolition waste in Hanoi, Journal of Cleaner Production, vol. 179 (2018), pp. 593-604.
[18] S. Moon: Studies on various structural system design options for twisted tall buildings and their performances, Structural Design of Tall and Special Buildings, Vol.23, no. 5 (2014), pp.319-333.
[19] F. Pacheco-Torgal, and S. Jalali: Toxicity of building materials: a key issue in sustainable construction, International Journal of Sustainable Engineering, vol. 4, no. 3 (2011), pp. 281-287.
[20] P. Xincheng: Super-high-strength high performance concrete, Hoboken: Taylor and Francis (2012).
[21] M. Gere, and J. Goodno: Mechanics of materials, 8th edition, Stamford: Cengage Learning. (2013).
[22] A. Yates: Sustainable Infrastructure – Sustainable Buildings, ICE Publishing (2015), pp 121-134.
[23] K. Gharehbaghi, M. Georgy, and F. Rahmani: Composite High-Rise Structures: Structural Health Monitoring (SHM) and Case Studies, Materials and Technologies in Engineering (2018), pp 146-152
[24] C. Kilbert: Sustainable construction: Green building design and delivery, Hoboken, Wiley (2018).
[25] C. Hibbeler, and S. Vijay: Mechanics of materials, 9th edition, worldwide adaptation edition, SI edition, Singapore: Pearson Education South Asia Pty Ltd (2014).
[26] O. Akadiri: Understanding barriers affecting the selection of sustainable materials in building projects, Journal of Building Engineering, vol. 4 (2015), 86-93, Elsevier.
[27] M. Mamlouk, J. Zaniewski: Materials for civil and construction engineers, Third edition, Prentice Hall (2011).
[28] A. Ramdasi, and N. Tande: Lateral load analysis of High Rise Structures, International Journal of Engineering Science and Technology, Vol 6, Iss 6 (2014), pp 268-275.
[29] E. Simiu, and D. Yeo: Advances in the design of high-rise structures by the wind tunnel procedure: Conceptual framework, Wind and Structures, An International Journal, vol. 21, no. 5 (2015), pp. 489-503.
[30] R. Umadevi, G. Singhb, and A. Chourasiyac: Laboratory Investigation on Use of Fly Ash Plastic Waste Composite in Bituminous Concrete Mixtures, Procedia - Social and Behavioral Sciences, Vol. 104, no. 2 (2013), pp. 89-98.
[31] L. Xing: The Rise and Potential Peak of Cement Demand in the Urbanized World Report, 'CORNERSTONE, The official Journal of world's coal industry (2015).
[32] X. Yan, X. Ren, X. and J. Li: Compressive behaviour of concrete under different loading rates: Experimental analysis, Life-Cycle of Engineering Systems: Emphasis on Sustainable Civil Infrastructure - 5th International Symposium on Life-Cycle Engineering, IALCCE (2016), pp. 1799.
[33] K. Gharehbaghi, and K. McManus: TIS condition monitoring using ANN integration: an overview, Journal of Engineering, Design and Technology, vol. 17, issue 1 (2019), pp. 204-217.
[34] B. Succar, W. Sher, A. and Willams: Measuring BIM performance: Five metrics, Journal Architectural Engineering and Design Management, 8(2) (2012), pp 120-142.
[35] K. Gharehbaghi, and C. Scott-Young: Waste diminution in construction projects: Environmental predicaments, Earth and Environmental Science, Institute of Physics, United Kingdom, vol. 127, no. 1. (2018b) pp. 58-63.
[36] K. Gharehbaghi, F. Rahmani: Viscoelastic materials: innovation and development countenance, Solid State Phenomena - Key Engineering materials, Trans Tech Publications, Switzerland, vol. 266 (2017), pp 165-171.

[37] Gharehbaghi, K., (2017), "Modular High-rise Construction as an alternative Building System", International Journal of the Constructed Environment, 8(3), pp 15-25.

[38] Gharehbaghi, K., McManus K., and Robson, K., (2019), "Minimizing the environmental impacts of mega infrastructure projects: Australian public transport perspective", Journal of Engineering, Design and Technology, vol. 17, issue 4, pp. 736-746.

[39] C. Zhang, C. and L. Canning: Application of non-conventional materials in construction", Proceedings of Institution of Civil Engineers: Construction Materials, vol. 164, no. 4 (2011), pp. 165-172.

Mechanical and Post-Cracking Performance of Recycled Aggregate Concrete Incorporating Synthetic Fibers

Syed M S Kazmi*, Muhammad J Munir, Yufei Wu, and Indubhushan Patnaikuni

School of Engineering, RMIT University, 376-392 Swanston St, Melbourne, Victoria-3001, Australia.

* s3636562@student.rmit.edu.au

Abstract. Reuse of concrete waste in novel construction is becoming very important topic nowadays. This study focuses to examine the post-cracking and mechanical performance, fracture behavior, and micro-structure of fiber strengthened recycled aggregate concrete (RAC). For this purpose, crack mouth opening diameter (CMOD) tests were conducted on twenty-seven notched beam specimens (550 x 150 x 150 mm) having three replacement levels (i.e., 0, 50 and 100%) of recycled concrete aggregates (RCA) and three synthetic fiber dosages (i.e., 0, 0.5 and 1%). Different mechanical properties of all mixes were also examined following ASTM standards. Drop in the mechanical performance of RAC was noticed at higher RCA replacement levels. However, synthetic fiber reinforced RAC showed better performance as compared to plain RAC. Results also depict positive influence of synthetic fiber addition on the residual flexural tensile strength of concrete. Approximately, 129% and 380% rise in toughness index and fracture energy was also observed for 1% fiber incorporation in RAC. Moreover, scanning electron microscopic analysis also confirmed the synthetic fiber-mortar bond. Therefore, synthetic fibers enhance the post-cracking and mechanical performance of fiber reinforced RAC resulting into more ductile and energy absorbing sustainable concrete.

1. Introduction

Concrete is widely utilized in different construction projects around the globe, which is produced after using a huge quantity of natural resources. Concrete aggregates constitute around 60-75% of all the volume of concrete [1]. Annual consumption of natural coarse aggregates (NCA) in concrete production is about 20 billion tons and is predicted to double itself in the next 30 years [1]. On the other hand, huge amount of waste is produced from construction and demolition (C&D) activities worldwide. For instance, China and Australia produce 200 Mt [2] and 13.7 Mt [3] C&D waste each year. Similarly, US, Europe and Japan produce 900 Mt of construction and demolition waste per year [4]. Best possible solution of the problem is to reuse this construction and demolition waste to yield recycled concrete aggregates (RCA). Consequently, landfilling problems as well as exhaustion of natural resources may be reduced to a great volume.

A lot of studies are available in the open literature regarding the utilization of different wastes in the novel construction materials [5-15]. Recycling of C&D waste in the form of RCA to produce recycled aggregate concrete (RAC) have also been investigated by many researchers in the past [1, 4, 7, 16, 17]. Reduction in mechanical performance of recycled aggregate concrete (RAC) was reported in previous studies due to inferior properties of RCA [18, 19]. Approximately, 25%, 23%and 24% decrease in the compressive, flexural strength and split tensile strength was observed in fully RCA incorporated RAC [1, 7, 19]. Variety of RCA treatment techniques have been proposed by researchers to improve RAC properties, including surface modification of RCA through carbonation or other techniques [20], removing adhered mortar from RCA [17], modifying RAC mixing method [21-23] ,utilization of contentious materials and incorporation of different fibers in RAC fibers [24-26]. Increase in the mechanical properties was reported in the fiber reinforced RAC. For example, 13%, rise in the compressive strength was observed for 0.75% steel fiber dosage [27]. Similarly, 4 and 31% surge in the flexural strength and split tensile strength was also reported after incorporating 1% synthetic fibers [28]. Moreover, improved flexural performance and impact resistance of fiber strengthened concrete was also reported due to utilization of synthetic fibers [29]

Nowadays different types of synthetic and steel fibers are utilized in variety of construction projects. Synthetic fibers are used in different types of constructions for example tunnel linings and industrial pavements due to lighter in weight, cheaper in cost, better distribution in concrete, improved impact resistance and durability as compared to steel fibers [16, 30]. Synthetic fibers have been reported very useful for normal concrete due to the associated better performance of resulting fiber reinforced concrete. However, there is a scant literature available on the effect of synthetic fibers on the properties of RAC. Consequently, more research is required to explore the mechanical and post-cracking properties of synthetic-fiber strengthened RAC. The aim of this study is to explore the post-cracking and mechanical properties of synthetic-fiber incorporated RAC. For this reason, nine different concrete mixes were prepared and tested with 0, 50 and 100% RCA replacement and 0, 0.5 and 1% synthetic fiber dosage.

2. Materials and mixture proportion

NCA and RCA having maximum size 14 mm were utilized in this research. RCA was attained from local concrete crushing plant having bulk density and water absorption 1413 kg/m^3 and 7.61% respectively. NCA comprised of crushed basalt aggregates having bulk density and water absorption 1512 kg/m^3 and 2.06% respectively. The gradation of RCA and NCA is depicted in figure. 1.

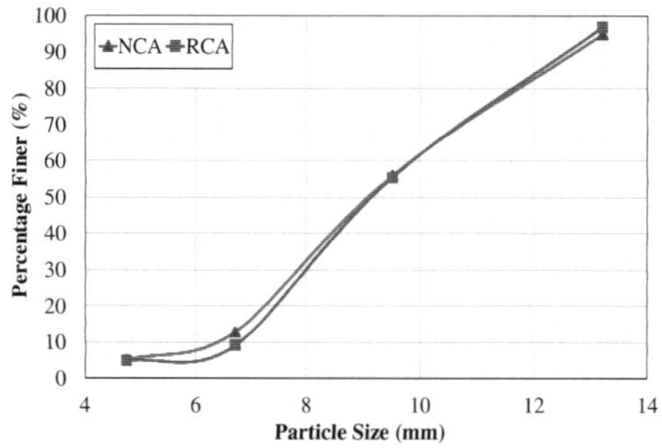

Figure 1. Gradation of NCA and RCA.

Figure 2. Polypropylene synthetic fibers.

RCA used in this study had less than 1% 30% contaminants and brick pieces respectively. Therefore, RCA can be categorized under class 1B [31-33]. Local sand from river source having specific gravity value of 2.65 and Ordinary Portland cement (OPC) were used throughout the study. Polypropylene

synthetic fibers were used in this study had length, equivalent diameter, young's modulus and density 57 mm 0.65 mm and 14 GPa and 0.91 kg/m3 respectively (figure. 2).
A total of nine concrete mixes with three RCA replacement levels (0,50 and 100%) and three fiber dosages (0, 0.5, 1%) were prepared and tested to investigate the performance of synthetic fiber reinforced RAC (table 1). Oven dried aggregates were used in this study and additional amount of water was mixed to concrete mixes depending upon the water absorption capacity of the aggregates.

3. Sample preparation and test methods

Concrete mixtures were prepared using pan mixer as per ASTM C192 / C192M - 16a [34]. Three concrete prisms having size 150 x 150 x 550 mm and six 150 mm diameter cylinders having 300 mm height were prepared for all the nine mixes. To evaluate the post-cracking performance of synthetic fiber strengthened RAC, a 25 mm high notch was made in the concrete prisms and tested under three point bending test to record load and Crack mouth opening displacement (CMOD) as per EN 14651:2005 [35]. Moreover, 28 days compressive strength and split tensile strength was also determined in accordance with ASTM C39:2017 [36] and ASTM C496:2017 [37], respectively.
Residual flexural tensile strength describes the post-cracking performance of the concrete sample was calculated at different CMOD values using Equation 1.

$$R_{,j} = \frac{3lF_j}{2bh_{sp}^2} \quad (1)$$

Where, Fj and Rj are the and load and residual flexural tensile strength at different CMOD (i.e. 0.5, 1.5, 2.5 and 3.5 mm). Similarly, b, l and Hsp are the width of beam, span of beam and the height of beam exclusive of the depth of notch, respectively. The fracture energy (Gf) of concrete samples was determined using Equation 2 [38].

$$G_f = \frac{W_0 + mg\delta}{A} \quad (2)$$

Where W0 denotes the area underneath the load-CMOD curve (N/m), mg is the beam weight among its supports (kg), δ is the peak deflection (m) and A is the area of fracture equivalent to [b(d - ao)] (m2), b, d and ao are the breadth, height of the notched beam specimen and height of notch, respectively. Toughness of all the fiber-reinforced samples were calculated using ratio of area underneath load-CMOD diagram of fiber-reinforced and plain concrete specimen [39]. Moreover, Philips XL30 scanning electron microscope (SEM) was used to investigate the micro-structure of fiber reinforced and plain concrete specimens.

4. Results and discussions

4.1 Mechanical performance of synthetic fiber reinforced concrete

Table 2 shows the average compressive and split tensile strength values of plain and synthetic-fiber strengthened normal aggregate concrete (NAC) and RAC concrete specimens. Drop in the compressive and split tensile strength values was noticed with rise in the RCA replacement percentage. For instance, 25% and 16% reduction of compressive and split tensile strength was noted for 100% RCA replacement. This reduction in mechanical performance of RAC may be attributed to the weak old mortar attached with the RCA surface [1, 40].
However, addition of synthetic fibers resulted in the improved mechanical properties of fiber-reinforced RAC. Addition of 1% synthetic fibers resulted in 5% and 30% rise in the compressive strength and split tensile strength of the RAC. Synthetic fibers improve the resistance of micro crack sliding [41] and help in transferring the stress from concrete to fibers [42] leading to improved mechanical performance of concrete. Similar improved mechanical properties of concrete due to fiber incorporation was also observed in past studies [25, 27].

Table 1. Mix proportion details.

Group ID	R0	R0P0.5	R0P1	R50	R50P0.5	R50P1	R100	R100P0.5	R100P1
Cement (kg/m^3)	451.4	451.4	451.4	451.4	451.4	451.4	451.4	451.4	451.4
Sand (kg/m^3)	708.6	708.6	708.6	708.6	708.6	708.6	708.6	708.6	708.6
NCA (kg/m^3)	822.5	822.5	822.5	411.2	411.2	411.2	---	---	---
RCA (kg/m^3)	---	---	---	411.2	411.2	411.2	822.5	822.5	822.5
Synthetic fibers (%)	---	0.5	1	---	0.5	1	---	0.5	1
Water (kg/m^3)	185.1	185.1	185.1	185.1	185.1	185.1	185.1	185.1	185.1
W/c	0.41	0.41	0.41	0.41	0.41	0.41	0.41	0.41	0.41
Extra Water (kg/m^3)	17.0	17.0	17.0	39.8	39.8	39.8	62.5	62.5	62.5
Slump (mm)	65	45	35	55	35	30	50	35	25

Table 2. Mechanical properties.

Mechanical properties	Compressive strength (MPa)	Split tensile strength (MPa)
R0	49.65	3.19
R0P0.5	50.20	3.80
R0P1	52.26	4.53
R50	41.48	3.24
R50P0.5	43.29	4.03
R50P1	44.24	4.39
R100	38.16	2.68
R100P0.5	40.16	3.69
R100P1	41.60	4.22

4.2 Post-cracking properties

Figures 3-5 depict the load-CMOD diagrams of all the concrete samples having three different dosages of synthetic fibers. It can be noticed that rise in RCA replacement level in the concrete results into lower peak load. For instance, 18% drop in the peak load was observed for 100% RCA replacement. Lower peak load may be due to weak old adhered mortar with RCA, similar results were reported in the past studies [31]. Steep and brittle post-peak performance was noted for the plain concrete specimens as revealed in figures 3-5. However, ductile behavior was observed in the samples incorporating synthetic fibers with steeper post-peak curve of RAC samples as equated to virgin concrete samples. For plain concrete samples, the test was conducted till failure. However, all the synthetic fiber reinforced samples were tested till 4 mm CMOD value as per [43]. Increasing trend of peak load was noticed with the rise in fiber amount for all the RCA replaced specimens (figure 4). For example, 20% rise in the peak load was noticed for 50% and 100% RCA replaced synthetic fiber strengthened concrete specimens as compared to the plain concrete specimens.

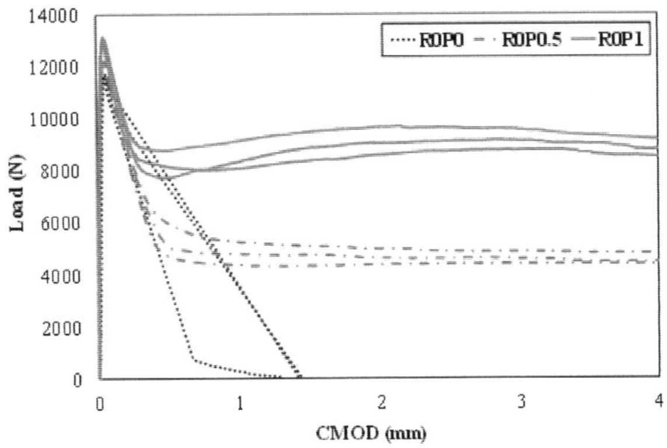

Figure 3. Effect of synthetic fibers on normal concrete.

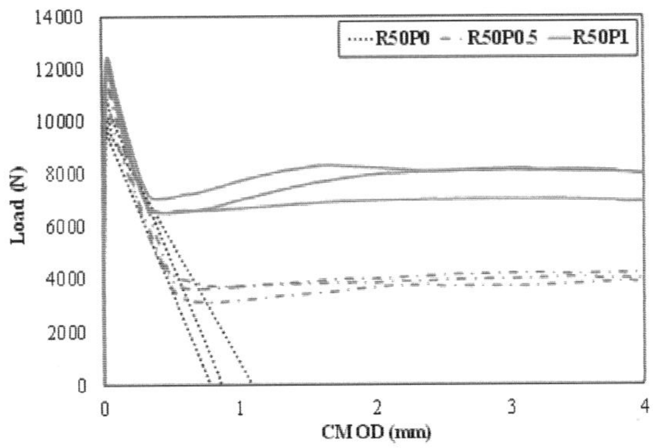

Figure 4. Effect of synthetic fibers on 50% RCA incorporated concrete.

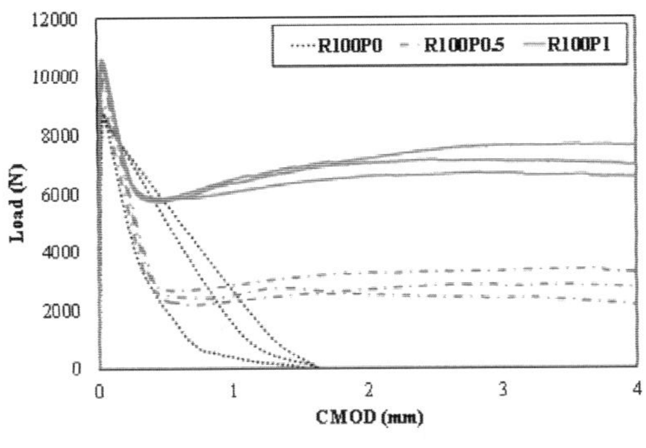

Figure 5. Effect of synthetic fibers on RAC.

Plain concrete specimens showed brittle type of failure. However, synthetic fiber reinforced specimens absorbed energy by bridging the cracks resulting in better post-peak behavior in the load-CMOD diagram. Moreover, through the descending branch of load-CMOD curve, synthetic fiber reinforced samples retrieved some of the load due to bond between concrete and fiber surface [44]. Comparable tendency was detected in past study [45]. Higher elongation potential, better bonding with the cracked concrete and medium tensile strength of synthetic fibers results into more energy absorption for large displacements leading towards better post cracking performance of concrete [46]. Therefore, synthetic fibers improve the ductility and post-peak behavior of resulting concrete specimens.

4.3 Fracture energy and toughness index

Figure 6 depicts the drop in the fracture energy of concrete samples with the rise in RCA replacement level. However, synthetic fibers improved the fracture energy of normal as well as RAC specimens. For example, around 22% drop in the fracture energy was observed by incorporating 100% RCA with the same fiber dosage (i.e. 1%). Moreover, 380 and 372% rise in the fracture energies was also noticed for NAC and RAC specimens with 1% synthetic fibers respectively. Surge in the fracture energy may be credited to de-bonding and stretching of synthetic fibers resulting in the prevention of crack propagation and higher absorption of energy [47].

Rise in the toughness value of concrete samples was noticed with the upsurge in synthetic fiber dosage (figure 7). For example, 123% rise in toughness was observed in the RAC specimens after the increase of synthetic fiber dosage from 0.5 to 1%. Moreover, RAC specimens depicted higher toughness index values as compared to normal concrete specimens. Similar increase in toughness values was also reported in past study [16].

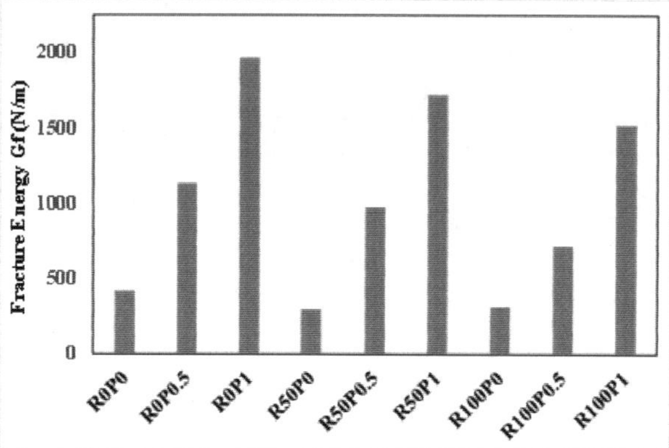

Figure 6. Fracture energy of concrete specimens.

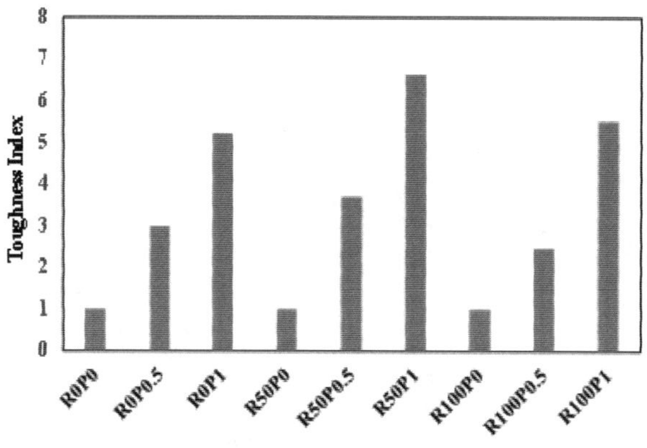

Figure 7. Toughness of concrete specimens.

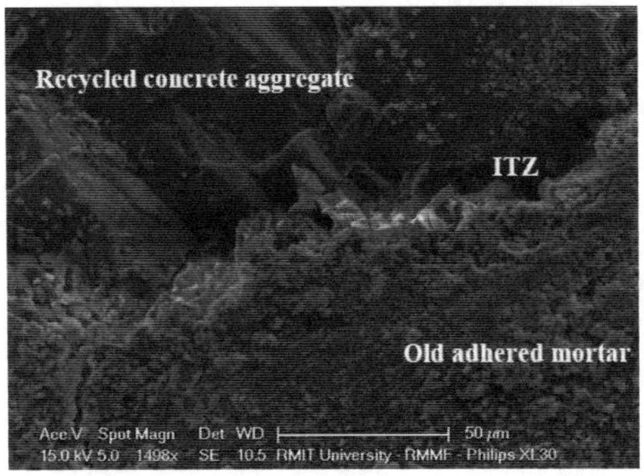

Figure 8. SEM image of RAC.

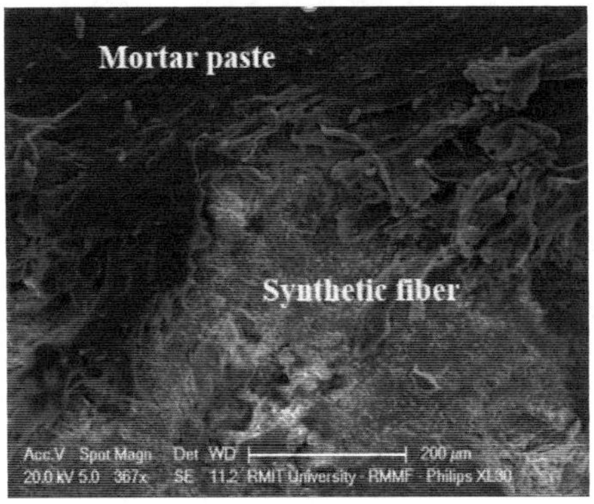

Figure 9. SEM image of fiber reinforced concrete.

4.4 Scanning electron microscopy

Figure 8 and figure 9 depict scanning electron microscope (SEM) images of RAC showing porous old adhered mortar attached with RCA and the microstructural interface of mortar and synthetic fibers in the concrete respectively. The lower post-cracking and mechanical performance of RAC in this study may be related to the weaker porous old adhered mortar attached with RCA. Figure. 9 also presents the rough surface of concrete along with smooth surface of synthetic fiber resulting in stronger fiber-matrix bond. Better fiber-matrix bond also resulted in improved post-cracking and mechanical properties of fiber-reinforced concrete as equated to plain concrete.

5. Summary and conclusions

Following conclusions were drawn based on this study

- Drop in the mechanical properties was observed with the RCA incorporation. However, synthetic fibers improved the mechanical performance of RAC.
- Synthetic fibers increased the highest load in load-CMOD diagram for normal as well as RAC specimens. RCA incorporation in concrete reduced the peak load of concrete specimens. Moreover, plain concrete specimens showed brittle post-peak behavior as compared to fiber reinforced specimens.
- Residual-flexural tensile strength of the concrete specimens increased with upsurge in synthetic fiber dosage resulting in improved post-cracking performance of normal concrete as well as RAC.
- RAC strengthened with 1% synthetic-fibers exhibited rise in the toughness and fracture energy by 129 and 380% respectively.
- SEM examination of displayed the porous old adhered mortar attached with the RCA, which caused inferior concrete performance. Presence of synthetic fiber-concrete bond was also noticed, which enhanced the post cracking and mechanical performance of fiber-reinforced concrete.

6. References

[1] Behera M, Bhattacharyya S K, Minocha A K, Deoliya R and Maiti S 2014 Recycled aggregate from C&D waste & its use in concrete – A breakthrough towards sustainability in construction sector: A review *Constr Build Mater.* **68** 501-16
[2] Productivity Commission 2006 Waste management: productivity commission draft report. (Canberra, Australia: Australian Government)
[3] Xiao J, Li W and Poon C 2012 Recent studies on mechanical properties of recycled aggregate concrete in China—A review *Sci. China Technol. Sci.* **55** 1463-80
[4] Oikonomou N D 2005 Recycled concrete aggregates *Cem Concr Compos.* **27** 315-8
[5] Kazmi S M S, Abbas S, Saleem M A, Munir M J and Khitab A 2016 Manufacturing of sustainable clay bricks: Utilization of waste sugarcane bagasse and rice husk ashes *Constr Build Mater.* **120** 29-41
[6] Syed Minhaj Saleem Kazmi M J M, Safeer Abbas, Muhammad Azhar Saleem, Anwar Khitab, Muhammad Rizwan 2017 Development of lighter and eco-friendly burnt clay bricks incorporating sugarcane bagasse ash *Pakistan J Eng App Sci.* **21** 1-5
[7] Kazmi S M S, Munir M J, Wu Y-F and Patnaikuni I 2018 Effect of macro-synthetic fibers on the fracture energy and mechanical behavior of recycled aggregate concrete *Constr Build Mater.* **189** 857-68
[8] Munir Muhammad J, Abbas S, Nehdi Moncef L, Kazmi Syed M S and Khitab A 2018 Development of Eco-Friendly Fired Clay Bricks Incorporating Recycled Marble Powder *J Mat Civil Eng.* **30** 04018069
[9] Munir M J, Kazmi S M S, Khitab A and Hassan M 2016 Utilization of rice husk ash to mitigate alkali silica reaction in concrete. In: *2nd International Multi-Disciplinary Conference (IMDC 2016),* (University of Lahore (Gujrat Campus)
[10] Kazmi S M S, Munir M J, Wu Y-F, Hanif A and Patnaikuni I Thermal performance evaluation of eco-friendly bricks incorporating waste glass sludge *J Clean Prod.* **172** 1867-80
[11] Munir M J, Wu Y-F, Kazmi S M S, Patnaikuni I, Zhou Y and Xing F 2019 Stress-strain behavior of spirally confined recycled aggregate concrete: An approach towards sustainable design *Res, Cons Rec.* **146** 127-39
[12] Kazmi S Minhaj S, Abbas S, Munir M J and Khitab A 2016 Exploratory study on the effect of waste rice husk and sugarcane bagasse ashes in burnt clay bricks *J Build Eng.* **7** 372-8
[13] Abbas S, Saleem M A, Kazmi S M S and Munir M J 2017 Production of sustainable clay bricks using waste fly ash: Mechanical and durability properties *J Build Eng.* **14** 7-14
[14] Munir M J, Kazmi S M S and Wu Y-F 2017 Efficiency of waste marble powder in controlling alkali–silica reaction of concrete: A sustainable approach *Constr Build Mater.* **154** 590-9
[15] Kazmi S M S, Munir M J, Patnaikuni I, Wu Y-F and Fawad U 2018 Thermal performance enhancement of eco-friendly bricks incorporating agro-wastes *Energy Build.* **158** 1117-29
[16] Buratti N, Mazzotti C and Savoia M 2011 Post-cracking behaviour of steel and macro-synthetic fibre-reinforced concretes *Constr Build Mater.* **25** 2713-22
[17] Kim Y, Hanif A, Kazmi S M S, Munir M J and Park C 2018 Properties enhancement of recycled aggregate concrete through pretreatment of coarse aggregates – Comparative assessment of assorted techniques *J. Clean. Prod.* **191** 339-49
[18] Kisku N, Joshi H, Ansari M, Panda S K, Nayak S and Dutta S C 2017 A critical review and assessment for usage of recycled aggregate as sustainable construction material *Constr Build Mater.* **131** 721-40
[19] Etxeberria M, Vázquez E, Marí A and Barra M 2007 Influence of amount of recycled coarse aggregates and production process on properties of recycled aggregate concrete *Cem Concr Res.* **37** 735-42
[20] Zhang J, Shi C, Li Y, Pan X, Poon C-S and Xie Z 2015 Performance Enhancement of Recycled Concrete Aggregates through Carbonation *J Matr Civil Eng.* **27** 04015029
[21] Tam V W Y and Tam C M 2008 Diversifying two-stage mixing approach (TSMA) for recycled aggregate concrete: TSMAs and TSMAsc *Constr Build Mater.* **22** 2068-77

[22] Tam V W Y, Tam C M and Le K N 2007 Removal of cement mortar remains from recycled aggregate using pre-soaking approaches *Resour Conserv Recycl.* **50** 82-101
[23] Tam V W Y, Tam C M and Wang Y 2007 Optimization on proportion for recycled aggregate in concrete using two-stage mixing approach *Constr Build Mater.* **21** 1928-39
[24] Dong J F, Wang Q Y and Guan Z W 2017 Material properties of basalt fibre reinforced concrete made with recycled earthquake waste *Constr Build Mater.* **130** 241-51
[25] Gao D, Zhang L and Nokken M 2017 Compressive behavior of steel fiber reinforced recycled coarse aggregate concrete designed with equivalent cubic compressive strength *Constr Build Mater.* **141** 235-44
[26] Gao D, Zhang L and Nokken M 2017 Mechanical behavior of recycled coarse aggregate concrete reinforced with steel fibers under direct shear *Cem Concr Compos.* **79** 1-8
[27] Carneiro J A, Lima P R L, Leite M B and Toledo Filho R D 2014 Compressive stress–strain behavior of steel fiber reinforced-recycled aggregate concrete *Cem Concr Compos.* **46** 65-72
[28] Akça K R, Çakır Ö and İpek M 2015 Properties of polypropylene fiber reinforced concrete using recycled aggregates *Constr Build Mater.* **98** 620-30
[29] Behfarnia K and Behravan A 2014 Application of high performance polypropylene fibers in concrete lining of water tunnels *Mater. Des.* **55** 274-9
[30] Banthia N and Sappakittipakorn M 2007 Toughness enhancement in steel fiber reinforced concrete through fiber hybridization *Cem Concr Res.* **37** 1366-72
[31] Commonwealth Scientific and Industrial Research Organization 1998 Guidance on the preparation of non-structural concrete made from recycled concrete aggregate. (Australia: Commonwealth Scientific and Industrial Research Organization)
[32] Tam V W-Y, Gao X-F and Tam C M 2006 Comparing performance of modified two-stage mixing approach for producing recycled aggregate concrete *Mag Con Res.* **58** 477-84
[33] Kazmi S M S, Munir M J, Wu Y-F, Patnaikuni I, Zhou Y and Xing F 2019 Influence of different treatment methods on the mechanical behavior of recycled aggregate concrete: A comparative study *Cem Concr Compos.* **104** 103398
[34] ASTM C192 / C192M - 16a 2016 Standard Practice for Making and Curing Concrete Test Specimens in the Laboratory. (West Conshohocken, PA: American Society of Testing and Materials)
[35] EN 14651:2005 2005 Test method for metallic fibered concrete - Measuring the flexural tensile strength (limit of proportionality (LOP), residual). (Brussels, Belgium: European Committee for Standardization)
[36] ASTM C39:2017 2017 Standard test method for compressive strength of cylindrical concrete specimens. (West Conshohocken, PA: American Society of Testing and Materials)
[37] ASTM C496:2017 2017 Standard test method for splitting tensile strength of cylindrical concrete specimens. (West Conshohocken, PA: American Society of Testing and Materials)
[38] Recommendation R D 1985 Determination of the Fracture Energy of Mortar and Concrete by Means of Three-Point Bend Tests on Notched Beames *Mater. Str.* **18** 285-90
[39] Gopalaratnam V S and Gettu R 1995 On the characterization of flexural toughness in fiber reinforced concretes *Cem Concr Compos.* **17** 239-54
[40] Kazmi S M S, Munir M J, Wu Y-F, Patnaikuni I, Zhou Y and Xing F 2019 Axial stress-strain behavior of macro-synthetic fiber reinforced recycled aggregate concrete *Cem Concr Compos.* **97** 341-56
[41] Li V C 1992 A simplified micromechanical model of compressive strength of fiber-reinforced cementitious composites *Cem Concr Compos.* **14** 131-41
[42] Mohammadhosseini H, Abdul Awal A S M and Mohd Yatim J B 2017 The impact resistance and mechanical properties of concrete reinforced with waste polypropylene carpet fibres *Constr Build Mater.* **143** 147-57
[43] 14651:2005 E 2005 Test method for metallic fibered concrete - Measuring the flexural tensile strength (limit of proportionality (LOP), residual). In: *European Committee for Standardization,* (Management Centre: rue de Stassart, 36 B-1050 Brussels

[44] Alberti M G, Enfedaque A and Gálvez J C 2014 On the mechanical properties and fracture behavior of polyolefin fiber-reinforced self-compacting concrete *Constr Build Mater.* **55** 274-88
[45] Alberti M, Enfedaque A and Gálvez J 2015 Comparison between polyolefin fibre reinforced vibrated conventional concrete and self-compacting concrete *Constr Build Mater.* **85** 182-94
[46] Alberti M, Enfedaque A, Gálvez J, Cánovas M and Osorio I 2014 Polyolefin fiber-reinforced concrete enhanced with steel-hooked fibers in low proportions *Mater. Des.* **60** 57-65
[47] Merta I and Tschegg E K 2013 Fracture energy of natural fibre reinforced concrete *Constr Build Mater.* **40** 991-7

IOP Conf. Series: Materials Science and Engineering **829** (2020) 012004 doi:10.1088/1757-899X/829/1/012004

Influence of Concrete Strength on the Stress-Strain Behavior of Spirally Confined Recycled Aggregate Concrete

Muhammad J Munir*, Syed M S Kazmi, Yufei Wu, and Indubushan Patnaikuni

School of Engineering, RMIT University, 376-392 Swanston St, Melbourne, Victoria-3001, Australia.

* s3636560@student.rmit.edu.au

Abstract. Costly and non-environment-friendly methods are used to improve the inferior behavior of recycled aggregate concrete (RAC). Conversely, the strength enrichment of concrete due to confinement provided by lateral reinforcement is ignored in the design of concrete compression members. The focus of this study is to investigate the role of pre-existing transverse reinforcement and different design strengths of concrete on the stress strain behavior of RAC. For this reason, stress-strain behavior of spiral steel confined concrete specimens having variable confinement pressure, recycled aggregates (RA) replacement percentage and design strength is investigated. The results show a drop in compressive strength of concrete with the increase in replacement percentage of RA. However, steel confinement has a positive role to counterbalance the adverse effect of RA replacement on concrete strength. Improved ductility and stress-strain behavior of RAC are observed with the increase in confinement pressure. Based on the results, the un-utilized pre-existing steel spiral reinforcement in the concrete compression members can offset the inferior performance of RAC resulting into sustainable and cost-effective construction.

1. Introduction

Demolishing of the older buildings and construction of new structures cause a huge quantity of construction and demolition (C&D) waste generation each year around the globe. For instance, China produces 0.2 billion tons of C&D waste yearly [1]. Similarly, Australia dumps roughly 5 million tons of waste concrete to landfills [2]. Concrete as a most widely used construction material worldwide is mainly composed of natural resources. Natural aggregates (NA) occupy nearly 50% of the total volume of concrete [3]. Recycling of C&D waste in the concrete as recycled aggregates (RA) to manufacture recycled aggregate concrete (RAC) can serve two functions i.e. natural resource conservation and saving of land fill spaces.

A lot of researchers have studied the utilization of different kinds of wastes in the novel construction materials to achieve sustainability in construction [4-7]. For instance, mechanical and durability performance of RAC has been studied to investigate the feasibility of recycling C&D waste in the form of RA [8-14]. Old adhered mortar attached on the surface of RA is the major difference between NA and RA resulting into the inferior performance of RAC as compared to the traditional concrete [3, 15, 16]. Inferior performance of RAC is the reason behind lower practical application of RAC (i.e. non-structural purposes only such as road bases). About, 20 and 40% reduction in the compressive strength and modulus of elasticity was stated for 100% RA replacement in RAC as compared to normal aggregate concrete (NAC) [17, 18]. Therefore, the performance of RAC deteriorates with the increase amount of RA replacement, which is the major worry regarding the utilization of RA in the structural concrete. Numerous investigators have explored variety of methods to improve the performance of RAC [6, 19, 20].

Lateral confinement using the materials like fiber-reinforced polymers (FRP) and steel tubes improves the compressive strength and other mechanical properties of NAC [21-23]. Similarly, performance of RAC can also be improved using confinement methods [24]. For example, Yang and Han [25] used steel tubes to confine RAC resulting into the improvement in the strength and deformation capacity. Compressive strength, ductility and stress strain behavior of NAC can also be improved by providing confinement using spiral steel reinforcement [26]. Limited work is available in the open literature regarding the role of lateral steel spirals in the enhancement of mechanical and stress strain performance of RAC. Lateral confinement in the form of steel spirals is commonly provided in the design of concrete columns. Contribution of these lateral steel spirals in the enhancement of compressive behavior of concrete columns is generally ignored in the current design method of compression members. It is therefore important to study the ability of un-used spiral steel reinforcement to offset the inferior performance of RAC. Steel spirals are already present in the concrete compression members. Therefore, performance enhancement of RAC through pre-existing confinement by steel spirals may be an economical, sustainable and easy to adopt solution for the structural application of RAC. The focus of this study is to investigate the role of pre-existing transverse reinforcement and different design strengths of concrete on the stress strain behavior of RAC. For this reason, stress-strain behavior of spiral steel confined concrete specimens having variable confinement pressure, recycled aggregates (RA) replacement percentage and design strength is investigated.

2. Materials and methods

RA and crushed granite NA having size 4 – 20 mm were used in this study. Figure 1 shows the size distribution curves for NA and RA fulfilling the limits of ASTM C33:2016 [27]. Table 1 shows the higher water absorption and lower bulk density of RA as compared to NA, which may be related to the presence of old adhered mortar with the surface of RA [28, 29]. Ordinary Portland cement (type P.II52.5R), river sand and tap water were used to manufacture all the mixes shown in table 2. Before concrete mixing, all the aggregates were first oven-dried at 105oC for 24 hours and then additional water was added in each mixture to consider the water absorption capacity of aggregates. Slump value between 90-100 mm was observed for series A and B. Similarly, 50-60 mm slump was noted for the concrete mixes belonging to series C.

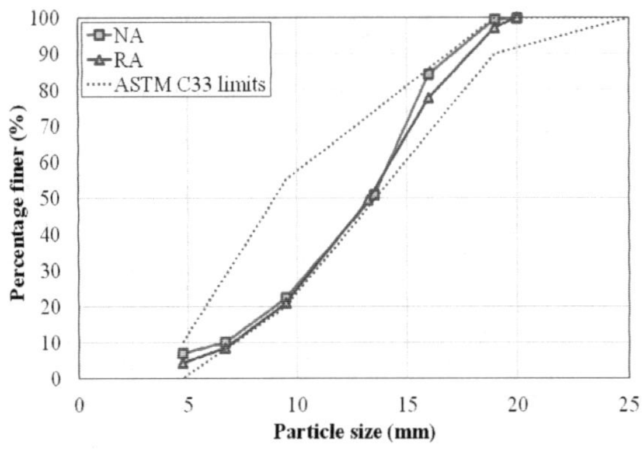

Figure 1. Gradation of NA and RA.

Table 1. Physical characteristics of NA and RA.

Property of aggregates	NA	RA
Water absorption (%)	1.30	6.85
Apparent specific gravity	2.66	2.55
Density (kg/m^3)	1513	1414
Crushing value (%)	27.0	31.0

To confine the concrete specimens, 3 mm round steel wire having yield and ultimate strength of 730 and 861 MPa, respectively was used as a spiral reinforcement. Elastic modulus and ultimate strain of steel wire were 200 GPa and 7.72%, respectively. For each series in table 2, forty-eight (i.e., 12 unconfined and 36 steel spiral confined) concrete specimens of size 150 x 300 mm were cast in accordance with ASTM C192:2016 [30] and 28 days curing was performed. Test variables included confinement pressures, replacement ratios of RA and target strengths of NAC specimens. Four different replacement ratios of RA (i.e., 0, 20, 50 and 100%) along with three different target strengths of NAC specimens (i.e., 25 MPa, 40 MPa and 60 MPa) were considered during the study. To consider varying confinement levels, three different pitches (i.e., center to center spacings) of steel spirals (i.e., 40, 30 and 20 mm, respectively) were selected considering Wu and Wei [31] model. To examine the pure effect of confinement pressure, no longitudinal reinforcement and concrete cover were considered in this study.

Table 2. Details of concrete mix design.

Series	RA replacement ratio (%)	Cement	Sand	NA	RA	Water	Extra water	Admixture (ml/m^3)
A	0	201.6	462.5	865.1	---	100.8	11.2	---
	20	201.6	462.5	691.5	161.6	100.8	19.9	---
	50	201.6	462.5	432.2	404.0	100.8	33.0	---
	100	201.6	462.5	---	807.9	100.8	55.3	---
B	0	288.0	528.8	710.1	---	135.3	9.2	---
	20	288.0	528.8	568.1	132.7	135.3	16.4	---
	50	288.0	528.8	355.0	331.8	135.3	27.3	---
	100	288.0	528.8	---	663.7	135.3	45.4	---
C	0	313.6	539.8	672.4	---	78.4	8.7	967.8
	20	313.6	539.8	537.9	125.6	78.4	15.6	966.7
	50	313.6	539.8	336.2	314.2	78.4	25.9	964.9
	100	313.6	539.8	---	628.4	78.4	43.0	962.0

For each mix combination, three specimens were cast, cured and tested. The labels of concrete cylinders display RA replacement percentage (R0, R20, R50 and R100 specify 0, 20, 50 and 100% RA replacement, respectively), pitches of steel spirals (S20, S30 and S40 mean pitch of 20, 30 and 40 mm, respectively) and design strengths of NAC (series A, B and C show the mix design with target strength of 25 MPa, 40 MPa and 60 MPa for NAC, respectively). For instance, R20-S40-B denotes the concrete mix with 20% of RA replaced with NA, 40 mm pitch of steel spirals and having the design strength of 40 MPa for NAC. Compression tests were performed on the specimens using displacement control mode with a rate of 0.3 mm/min. Data acquisition system and four linear variable displacement transducers (LVDTs) mounted at 90o relative to each other were used to record the load and to measure the axial deformation, respectively. During the compression test, applied load and deformation were measured to examine the stress strain behavior of all the concrete specimens.

3. Results and discussions

IOP Conf. Series: Materials Science and Engineering **829** (2020) 012004 doi:10.1088/1757-899X/829/1/012004

3.1 Axial stress strain performance

Figure 2 depicts the effect of RA replacement on the stress strain behavior of all the confined and unconfined specimens. The axial strain of all the concrete specimens was calculated by the average value from the four LVDTs mounted vertically around the specimen. Shape of the stress-strain curves for all the confined and un-confined specimens changes steadily and flatten with the increase in RA replacement ratio. Reduction in the elastic modulus (i.e. initial slope) and a lower peak load are observed for the un-confined concrete specimens with the rise in RA replacement ratio as shown in figures 2(a-c). For series-A, un-confined concrete specimens display a reduction in the slope of descending branches of stress strain curves with the rise in replacement ratio of RA. This may be attributed to the inferior strength of concrete in series-A with the increase in RA replacement ratio. For the series B and C, the descending portions of the stress-strain curves are observed steeper with the rise in RA replacement ratio, which may be attributed to the more brittle behavior of RAC in comparison with NAC as reported in a past studies [32, 33]. Un-confined concrete specimens also show reduced energy absorption capacity because of the smaller area under the curve with the rise in RA replacement ratio.

For all the confined specimens, lower stress-strain curves are obtained with the rise in RA replacement ratio, as shown in figures 2(d-l). Effect of confinement in terms of strength improvement increases with the rise in the replacement ratio of RA. Furthermore, the effect of confinement is observed more prominent for lower strength concrete such as series-A concrete specimens as compared to series B and C specimens. Similar findings are reported in the previous study by Wang et al. [34]. Reduction in the compressive strength due to the increase in RA replacement ratio is also observed in the presence of lateral confinement, which is consistent with the observations of Teng et al. [35] for FRP confined RAC.

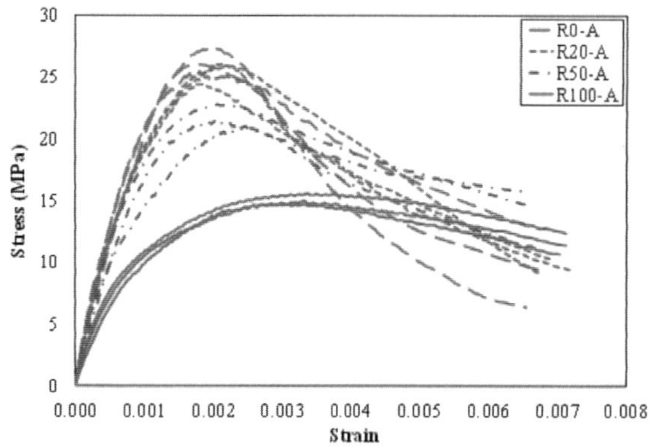

(a)

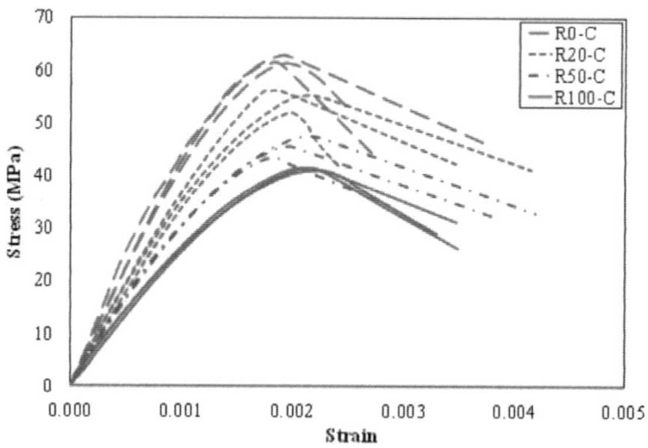

(c)

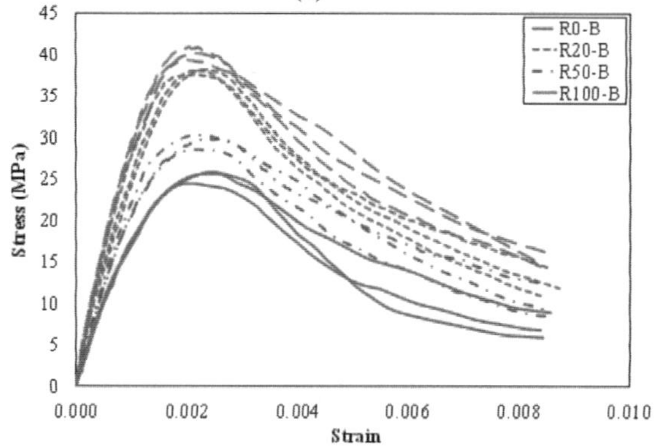

(b)

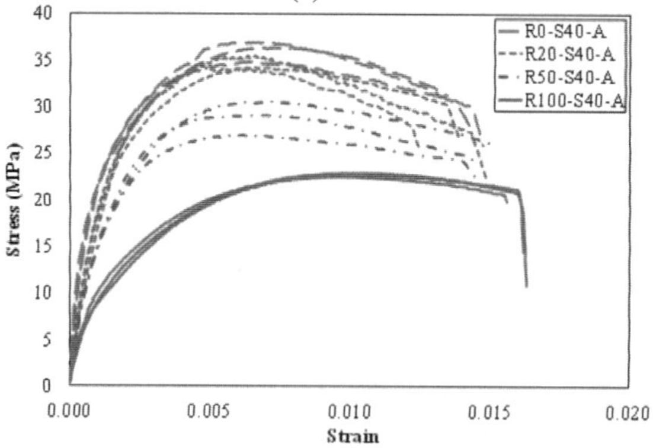

(d)

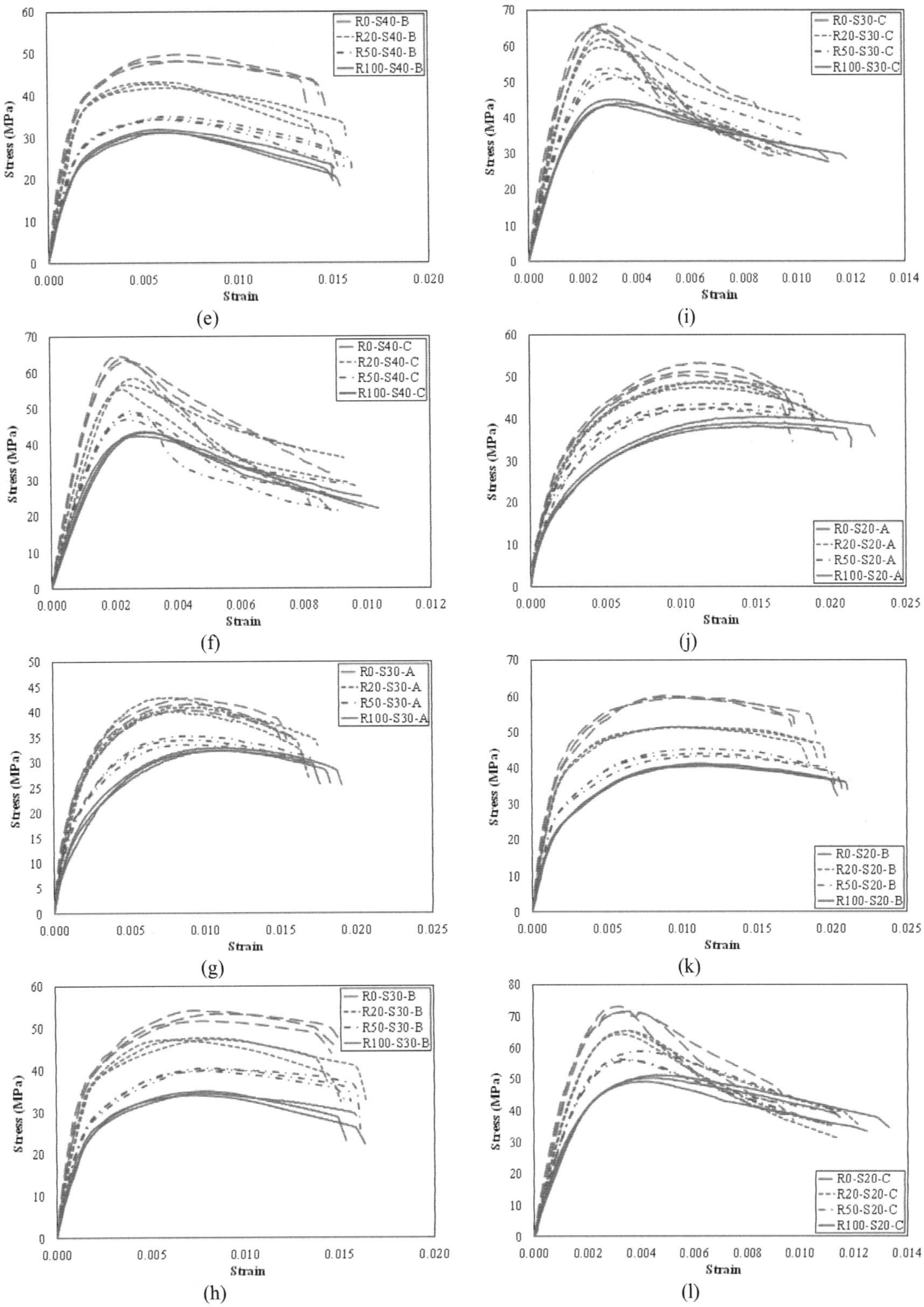

Figure 2. Effect of RA replacement ratio on the stress-strain behaviour of (a) unconfined specimens – Series-A, (b) unconfined specimens – Series-B, (c) unconfined specimens – Series-C, (d) confined S40 specimens – Series-A, (e) confined S40 specimens – Series-B, (f) confined S40 specimens – Series-C, (g) confined S30

specimens – Series-A, (h) confined S30 specimens – Series-B, (i) confined S30 specimens – Series-C, (j) confined S20 specimens – Series-A, (k) confined S20 specimens – Series-B, and (l) confined S20 specimens – Series-C.

3.2 *Compressive strength*

Table 3 shows the effect of strength, RA replacement ratio and confinement pressure on the average compressive strength values of all the confined and un-confined concrete specimens. For all the specimens, the drop in compressive strength is observed with the increase in RA content. For instance, R100-A, R100-B and R100-C specimens show 43, 37 and 33% decline in the average compressive strength as compared to R0-A, R0-B and R0-C specimens, respectively. The difference in strength drop is linked to the difference in concrete mix design for each series. Similar trend was also reported by Pacheco et al. [36]. The increase in the confinement pressure results into the increase in the average compressive strength of the confined concrete specimens. All the confined concrete specimens of series-A except R100-S40-A show average compressive strength higher as compared to un-confined R0-A specimens. Similarly, all the confined concrete specimens of series-B except R50-S40-B, R100-S40-B and R100-S30-B display average strength higher as compared to the un-confined R0-B specimens. For series-C, the compressive strength values of R0-S40-C, R0-S30-C, R0-S20-C and R20-S20-C specimens are observed higher as compared to un-confined R0-C specimens, whereas, all the other specimens show compressive strength still lower than un-confined R0-C specimens. By comparing, R100-S40-B with R0-S40-B and R100-S20-B with R0-S20-B, drop in strength due to the use of RA reduces to 36% and 32% after providing the confinement with 40 mm and 20 mm pitch of spiral reinforcement, respectively. Similar behavior is observed for series A and C concrete specimens, which shows the positive role of confinement pressure to counter the adverse effects of RA replacement. Similar trend was also observed by Zhao et al. [37] for the FRP confined RAC specimens.

Table 3. Mechanical properties of all the specimens.

Specimen	Series A (25 MPa)		Series B (40 MPa)		Series C (60 MPa)	
	Compress-ive strength (MPa)	Modulus of elasticity (GPa)	Compress-ive strength (MPa)	Modulus of elasticity (GPa)	Compress-ive strength (MPa)	Modulus of elasticity (GPa)
R0	26.26	25.30	40.10	33.76	61.82	46.93
R20	25.46	19.89	37.95	29.08	54.36	37.25
R50	21.78	19.72	29.55	24.52	45.45	32.99
R100	15.09	12.82	25.31	20.30	41.30	27.67
R0-S40	36.05	25.78	48.63	33.16	63.88	46.55
R20-S40	34.66	19.85	42.48	25.68	56.70	37.96
R50-S40	28.88	16.93	34.56	21.09	48.31	32.59
R100-S40	22.75	12.72	31.34	18.71	43.07	27.09
R0-S30	41.56	22.99	53.05	31.16	65.22	46.18
R20-S30	41.41	18.71	47.37	27.13	61.72	38.05
R50-S30	34.37	16.68	40.05	21.95	52.50	33.39
R100-S30	32.51	12.08	34.36	18.50	44.22	27.77
R0-S20	51.63	19.32	59.67	28.69	71.95	42.79
R20-S20	48.33	18.09	51.22	24.28	64.99	35.61
R50-S20	42.82	16.70	44.09	19.39	57.09	31.86
R100-S20	39.25	11.91	40.78	16.62	50.22	26.95

3.3 *Modulus of elasticity*

Table 3 depicts the effect of design strength, RA replacement ratio and confinement pressure on the average elastic modulus of all the confined and un-confined concrete specimens. Modulus of elasticity of all the specimens was calculated using the initial slope of the axial stress-strain curve following the method used by Carneiro, Lima, Leite and Toledo Filho [38]. For all the specimens, the drop in the modulus of elastic is noticed with the increase in RA replacement ratio. For instance, R100-A, R100-B and R100-C specimens display 49, 40 and 41% decline in average modulus of elasticity as compared to R0-A, R0-B and R0-C specimens, respectively. The drop in modulus of elasticity of concrete specimens may be linked to the inferior modulus of elasticity of RA (owing to porous old attached mortar), whereas, the difference in the reduction of modulus of elasticity for each series is related to the difference in the concrete mix design. Similar trend of modulus of elasticity was also reported in the previous studies [39, 40]. The increase in confinement pressure results in a drop in average modulus of elasticity. For instance, R20-A specimens show modulus of elasticity of 19.9 GPa, which reduces to 19.8, 18.7 and 18.1 GPa for R20-S40-A, R20-S30-A and R20-S20-A specimens, respectively. Similarly, R100-B specimens show modulus of elasticity of 20.3 GPa, which decreases to 18.7, 18.5 and 16.6 GPa for R100-S40-B, R100-S30-B and R100-S20-B specimens, respectively. In past studies, researchers [41, 42] also reported the reduction in modulus of elasticity of NAC with the increase in confinement pressure. The drop in the modulus of elasticity of concrete is attributed to the presence of lateral steel spirals, which form gaps between the surrounding concrete and steel spirals owing to the shrinkage of concrete [43]. Reduction of the cross-sectional area of concrete at the location of steel spirals due to the concrete shrinkage may result in decrease in the effective concrete area leading towards drop in the modulus of elasticity of confined specimens.

4. Summary and conclusions

Costly and non-environment-friendly methods are used to improve the inferior behavior of recycled aggregate concrete (RAC). Conversely, the strength enrichment of concrete due to confinement provided by lateral reinforcement is ignored in the design of concrete compression members. The focus of this study is to investigate the role of pre-existing transverse reinforcement and different design strengths of concrete on the stress strain behavior of RAC. For this reason, stress-strain behavior of spiral steel confined concrete specimens having variable confinement pressure, recycled aggregates (RA) replacement percentage and design strength is investigated. Based on results obtained during this study, following conclusions may be drawn.

- Steel spirals are already present in the concrete compression members. Therefore, performance enhancement of RAC through pre-existing confinement by steel spirals may be an economical, sustainable and easy to adopt solution for the structural application of RAC.
- Shape of the stress-strain curves for all the confined and un-confined specimens changes steadily and flatten with the increase in RA replacement ratio. However, improved stress-strain behavior is observed in the confined concrete specimens.
- Effect of confinement in terms of strength improvement increases with the rise in the replacement ratio of RA. The effect of confinement is also observed more prominent for lower strength concrete such as series-A concrete specimens as compared to series B and C specimens.
- Confinement through spiral steel reinforcement can offset the inferior performance of RAC. For example, all the spiral steel confined concrete specimens of series A except R100-S40-A show average compressive strength higher in comparison with unconfined R0-A specimens.
- Modulus of elasticity for all the specimens reduces with the increase in the amount of RA. Moreover, decline in the modulus of elasticity of spiral steel confined specimens is also observed which may be related to the decrease in the effective concrete area due to the shrinkage of concrete at the location of the spiral.
- Based on the results, the pre-existing steel spiral reinforcement in the concrete compression members can offset the inferior performance of RAC resulting into sustainable and cost-effective construction.

5. References

[1] Xiao J, Li W and Poon C 2012 Recent studies on mechanical properties of recycled aggregate concrete in China—A review *Sci China Tech Sci.* **55** 1463-80
[2] Tam V W Y 2009 Comparing the implementation of concrete recycling in the Australian and Japanese construction industries *J. Clean Prod.* **17** 688-702
[3] Shi C, Li Y, Zhang J, Li W, Chong L and Xie Z 2016 Performance enhancement of recycled concrete aggregate – A review *J. Clean Prod.* **112** 466-72
[4] Kazmi S M S, Abbas S, Saleem M A, Munir M J and Khitab A 2016 Manufacturing of sustainable clay bricks: Utilization of waste sugarcane bagasse and rice husk ashes *Const Build Mat.* **120** 29-41
[5] Kazmi S M S, Munir M J, Abbas S, Saleem M A, Khitab A, Rizwan M 2017 Development of lighter and eco-friendly burnt clay bricks incorporating sugarcane bagasse ash *Pak J Eng & App Sci.* **21** 1-5
[6] Kazmi S M S, Munir M J, Wu Y-F and Patnaikuni I 2018 Effect of macro-synthetic fibers on the fracture energy and mechanical behavior of recycled aggregate concrete *Const Build Mat.* **189** 857-68
[7] Munir M J, Kazmi S M S and Wu Y-F 2017 Efficiency of waste marble powder in controlling alkali–silica reaction of concrete: A sustainable approach *Const Build Mat.* **154** 590-9
[8] Xuan D, Zhan B and Poon C S 2017 Durability of recycled aggregate concrete prepared with carbonated recycled concrete aggregates *Cem Con Comp.* **84** 214-21
[9] Oikonomou N D 2005 Recycled concrete aggregates *Cem Con Comp.* **27** 315-8

[10] Kazmi S Minhaj S, Abbas S, Munir M J and Khitab A 2016 Exploratory study on the effect of waste rice husk and sugarcane bagasse ashes in burnt clay bricks *J Build Eng.* **7** 372-8
[11] Abbas S, Saleem M A, Kazmi S M S and Munir M J 2017 Production of sustainable clay bricks using waste fly ash: Mechanical and durability properties *J Build Eng.* **14** 7-14
[12] Kazmi S M S, Abbas S, Nehdi M L, Saleem M A and Munir M J 2017 Feasibility of using waste glass sludge in production of ecofriendly clay bricks *J Mat Civil Eng.* **29**
[13] Kazmi S M S, Munir M J, Wu Y-F, Hanif A and Patnaikuni I 2018 Thermal performance evaluation of eco-friendly bricks incorporating waste glass sludge *J Clean Prod.* **172** 1867-80
[14] Kazmi S M S, Munir M J, Patnaikuni I, Wu Y-F and Fawad U 2018 Thermal performance enhancement of eco-friendly bricks incorporating agro-wastes *Energ Buildings* **158** 1117-29
[15] Sáez del Bosque I F, Zhu W, Howind T, Matías A, Sánchez de Rojas M I and Medina C 2017 Properties of interfacial transition zones (ITZs) in concrete containing recycled mixed aggregate *Cem Con Comp.* **81** 25-34
[16] de Brito J, Ferreira J, Pacheco J, Soares D and Guerreiro M 2016 Structural, material, mechanical and durability properties and behaviour of recycled aggregates concrete *J Build Eng.* **6** 1-16
[17] Xiao J, Li W, Fan Y and Huang X 2012 An overview of study on recycled aggregate concrete in China (1996–2011) *Const Build Mat.* **31** 364-83
[18] Kou S C, Poon C S and Chan D 2007 Influence of fly ash as cement replacement on the properties of recycled aggregate concrete *J Mat Civil Eng.* **19** 709-17
[19] Kazmi S M S, Munir M J, Wu Y-F, Patnaikuni I, Zhou Y and Xing F 2019 Axial stress-strain behavior of macro-synthetic fiber reinforced recycled aggregate concrete *Cem Con Comp.* **97** 341-56
[20] Kim Y, Hanif A, Kazmi S M S, Munir M J and Park C 2018 Properties enhancement of recycled aggregate concrete through pretreatment of coarse aggregates – Comparative assessment of assorted techniques *J Clean Prod.* **191** 339-49
[21] Ma C-K, Apandi N M, Sofrie C S Y, Ng J H, Lo W H, Awang A Z and Omar W 2017 Repair and rehabilitation of concrete structures using confinement: A review *Const Build Mat.* **133** 502-15
[22] Mander J B, Priestley M J N and Park R 1988 Theoretical stress-strain model for confined concrete *J Str Eng.* **114** 1804-26
[23] Mander J B, Priestley M J N and Park R 1988 Observed stress-strain behavior of confined concrete *J Str Eng.* **114** 1827-49
[24] Xu J-J, Chen Z-P, Xiao Y, Demartino C and Wang J-H 2017 Recycled aggregate concrete in FRP-confined columns: A review of experimental results *Comp Str.* **174** 277-91
[25] Yang Y-F and Han L-H 2006 Experimental behaviour of recycled aggregate concrete filled steel tubular columns *J Const Steel Res.* **62** 1310-24
[26] Ahmad S H and Shah S P 1982 Stress-strain curves of concrete confined by spiral reinforcement *ACI J.* **79** 484-90
[27] ASTM C33:2016 2016 Standard specification for concrete aggregates. In: *American Society for Testing and Materials,* West Conshohocken, PA
[28] Kou S-C and Poon C-S 2013 Long-term mechanical and durability properties of recycled aggregate concrete prepared with the incorporation of fly ash *Cem Con Comp.* **37** 12-9
[29] Kou S-C, Poon C-S and Etxeberria M 2011 Influence of recycled aggregates on long term mechanical properties and pore size distribution of concrete *Cem Con Comp.* **33** 286-91
[30] ASTM C192:2016 2016 Standard practice for making and curing concrete test specimens in the laboratory. In: *American Society for Testing and Materials,* West Conshohocken, PA
[31] Wu Y-F and Wei Y 2015 General stress-strain model for steel- and FRP-confined concrete *Journal of Comp Const.* **19** 10.1061/(ASCE)CC.943-5614.0000511
[32] Xiao J, Zhang K and Akbarnezhad A 2018 Variability of stress-strain relationship for recycled aggregate concrete under uniaxial compression loading *J Clean Prod.* **181** 753-71
[33] Kazmi S M S, Munir M J, Wu Y-F, Patnaikuni I, Zhou Y and Xing F 2019 Influence of different treatment methods on the mechanical behavior of recycled aggregate concrete: A comparative study *Cem Con Comp.* **104** 103398
[34] Wang W, Zhang M, Tang Y, Zhang X and Ding X 2017 Behaviour of high-strength concrete columns confined by spiral reinforcement under uniaxial compression *Const Build Mat.* **154** 496-503
[35] Teng J G, Zhao J L, Yu T, Li L J and Guo Y C 2016 Behavior of FRP-confined compound concrete containing recycled concrete lumps *J Comp Const.* **20** 10.1061/(ASCE)CC.943-5614.0000602

[36] Pacheco J, de Brito J, Chastre C and Evangelista L 2019 Experimental investigation on the variability of the main mechanical properties of concrete produced with coarse recycled concrete aggregates *Const Build Mat.* **201** 110-20
[37] Zhao J L, Yu T and Teng J G 2015 Stress-strain behavior of FRP-confined recycled aggregate concrete *J Comp Const.* **19** 10.1061/(ASCE)CC.943-5614.0000513
[38] Carneiro J A, Lima P R L, Leite M B and Toledo Filho R D 2014 Compressive stress–strain behavior of steel fiber reinforced-recycled aggregate concrete *Cem Con Comp.* **46** 65-72
[39] Xiao J, Li J and Zhang C 2005 Mechanical properties of recycled aggregate concrete under uniaxial loading *Cem Conc Res.* **35** 1187-94
[40] Counto U J 1964 The effect of the elastic modulus of the aggregate on the elastic modulus, creep and creep recovery of concrete *Mag Conc Res.* **16** 129-38
[41] Ren X, Liu K, Li J and Gao X 2017 Compressive behavior of stirrup-confined concrete under dynamic loading *Const Build Mat.* **154** 10-22
[42] Richart F, Brandtzaeg A and Brown R 1929 The failure of plain and spirally reinforced concrete in compression. In: *Engineering Experiment Station, Bulletin no. 190,* (Urbana, USA: University of Illinois)
[43] Munir M J, Wu Y-F, Kazmi S M S, Patnaikuni I, Zhou Y and Xing F 2019 Stress-strain behavior of spirally confined recycled aggregate concrete: An approach towards sustainable design *Res, Cons Rec.* **146** 127-39

Authors' background

Your Name	Title	Research Field	Personal website
Muhammad Junaid Munir	PhD candidate	Civil Engineering	https://scholar.google.ca/citations?user=K8i9ntQAAAAJ&hl=en
Syed Minhaj Saleem Kazmi	PhD candidate	Civil Engineering	https://scholar.google.com.pk/citations?user=riKyfDgAAAAJ&hl=en
Yufei Wu	Full professor	Civil Engineering	https://scholar.google.com/citations?user=8I0lD_UAAAAJ&hl=en&oi=sra
Indubhushan Patnaikuni	Senior lecturer	Civil Engineering	https://scholar.google.com/citations?hl=en&user=l2kkgbYAAAAJ&view_op=list_works&sortby=pubdate

IOP Conf. Series: Materials Science and Engineering **829** (2020) 012005 doi:10.1088/1757-899X/829/1/012005

Feasibility of PET Resin as a Cement Substitute for Sustainable Construction

Y D Patil[1] **S M Waysal**[2] and **B Z Dholakiya**[3]

[1]Associate Professor in Applied Mechanics Department, Sardar Vallabhbhai National Institute of Technology, Surat, Gujarat, India
[2]Assistant Profesoor in civil engineering department, MVP samja's KBT college of engineering, Nashik.
[3] Associate Professor in Applied Chemistry Department, Sardar Vallabhbhai National Institute of Technology, Surat, Gujarat, India

[1]yogeshdpatilsurat@gmail.com, [2]satish.waysal@gmail.com, [3]bharat281173@gmail.com

Abstract: Polyethylene Terephthalate (PET) is one of the most widely used plastics in the packaging industry. PET need hundreds of years to decompose. There is an increasing demand for cement in the construction industry. If we could use PET in construction, it will be an excellent relief for the environment.
In this study, PET bottles were used to produce a PET resin by glycolysis in polyethylene glycol. These resins were prepared for a different proportion of PET particles and were added in 5, 10, and 15 percent by mass of cement. The effect of this resin on standard consistency, setting time, together with slump flow and compressive strength of cement paste, is studied. It is observed that the addition of PET resin alters the standard consistency and compressive strength, increases the setting time and slump flow. The findings of this study suggest that PET waste can be utilized to replace cement.

1. Introduction

The production of plastic waste has been expanding considerably, and the majority of it is sent to dumping grounds. The enthusiasm for recycling it, both ecologic and economic, is not just because of its forcefulness towards the earth [1] yet in addition to its high volume in dumping grounds and high protection from atmospheric and biological agents, making this arrangement unsustainable.
The primary procedure of waste dispersal was biologically common. This has brought about expansive amounts of garbage in the landfill[2] that is not decreased because of the disposal of junk contrasted with a high rate of waste disposal. Therefore, the need to open another disposal expanded every once in a while. Polyethylene Terephthalate (PET) is a sort of plastic that is typically utilized for plastic bottles.[3] PET is utilized in carbonated drinks, mineral water and packaging industry in the whole world. PET has advantages such as high solidness high weight resistance, non-reactivity with substance, incredible nature of gas catching which can safeguard the gas in the beverages [4]. The construction industry uses cement on a large scale. About a few billion tons are being used every year. Utilizing PET resin in concrete as the polymer is an ideal route on account of depleting natural resources and sustainable development and having durable and strong concrete.[5]

In the last twenty years, the study has been carried out on the use of waste plastics as concrete and mortar constituent. The PET waste is utilized as replacement of constitutes of concrete such as fine aggregate and cement. [6] The PET fibre is used as a reinforcement in concrete. The polyethylene terephthalate (PET) is one of the widely consumed plastic, in the total plastic waste generated. Nowadays, the literature has shown that many applications on mortar and concrete with utilizing plastic waste especially with the use of PET waste.[7]. The inclusion of PET fibres in the concrete is advantageous in order to control plastic shrinkage cracking. The plastic shrinkage cracks are widely explicited in a mortar, which has a large surface of exposure.[8] when surfaces are subjected to wetting and drying or freeze-thawing plastic shrinkage cracks are developed resulted in the rapid ruin of structure.[9] PET fibre has a low elastic modulus which is sufficient to give high performance to restrain cracks propagation in a mortar.[10]. Fibres in mortar act as uniformly distributing reinforcing material against the crack enlargement due to plastic shrinkage. Propagation of microcracks and macrocracks is a severe issue in harden mortar and it can be arrested by providing uniform distribution of PET. PET fibres present in mix act as a bridge and stop the propagation of microcracks. Free shrinkage and restrained shrinkage tests have performed the evaluation of fibres capacity to reduce plastic shrinkage [11]. The diminish in free shrinkage does not necessarily indicate a reduction in overall cracking tendency. The cracking sensitivity of harden mix is primarily controlled by enhanced toughness owing to the incorporation of fibers and shrinkage deformation [12]. The toughness plays a very important role during the performance in static, dynamic and fatigue actions.
PET waste can be handled by three ways reuse, recycle and reduce. Reuse of PET waste bottle is not feasible as they are made for single use and need to be crushed after use. Recycling of PET waste may be carried out by two ways one is by the mechanical way (crushing) and second by changing its chemical structure converting as PET resin. Generation of PET waste can be reduced by consuming less in day to day life, but at present, there is no other material to substitute PET, so reduce is not a feasible option. In present situation recycling of PET waste is the best option.

2. Materials and methods

2.1 Cement
Ultra Tech 53 grades Ordinary Portland cement is used for this study the specific gravity of cement is 3.10

2.2 Fine aggregate
River sand is used as fine aggregate it has specific gravity 2.62, water absorption 1.10, and confirming to zone I of IS 383.

2.3 PET resin
PET Resin was obtained by glycolysis process with 5,10 and 15 percentages of PET (Polyethylene terephthalate) partials in PEG (Polyethylene Glycol). Glycolysis of waste PET was done using PEG having 100 molecular weight [13] along with 0.5% zinc acetate as a catalyst. A three-necked glass flask with condenser and stirrer was used to conduct the reaction. The ingredients were heated at 180°C for an hour; the temperature was further elevated till the disappearance of all the solids at 210°C [14]. Styrene monomer is added in PET resin with ratio 1:1 to reduce the viscosity and Benzoyl peroxide was added in this mixture as a curing agent [15]. Oven dry and normal water curing was carried out on the mortar block. Benzoyl peroxide requires higher temperature to act as a curing agent for PET resin.

2.4 Testing program
Standard consistency, Initial setting time and final setting time is determined as per IS 4031-1968 and IS 269-1976. Soundness is determined as per IS 4031-1968 and IS 269-1976. The mix ratio for normal cement mix is 1:3 (Cement: Sand) and water of quantity $(\frac{P}{4} + 3)$ percent of the combined weight of cement and sand mix chosen for determination of compressive strength. PET resin is added in normal

cement mix with 5%, 10%, and 15% by weight of cement for all PET to glycol ratio. The flow table test is carried out according to IS 4031-part IV 2005. The compressive strength of cement mortar is determined as per IS 4031-part IV 2005.

3. Results and discussion

3.1 Consistency
The water required for standard consistency of the cement was slightly reduced with an increase in the addition of PET resin in the cement paste. The water requirement is reduced up to 2% and 5% for 10% PET resin and 15% PET resin respectively the same is reflected in table no 1. There is no effect on consistency with an increase in PET to glycol content.

Table 1. Effect of PET resin percentage and PET to glycol ratio on consistency and soundness on cement paste.

Sr. No	PET to glycol ratio	Specimen	Consistency Pn %	Soundness in mm
1	0	Cement	30.5	2.00
2	5	5% Resin	30.5	1.00
3		10% Resin	30.0	1.00
4		15% Resin	29.0	1.00
5	10	5% Resin	30.5	1.00
6		10% Resin	29.0	1.50
7		15% Resin	28.0	2.00
8	15	5% Resin	30.0	1.00
9		10% Resin	29.0	2.00
10		15% Resin	28.0	2.00

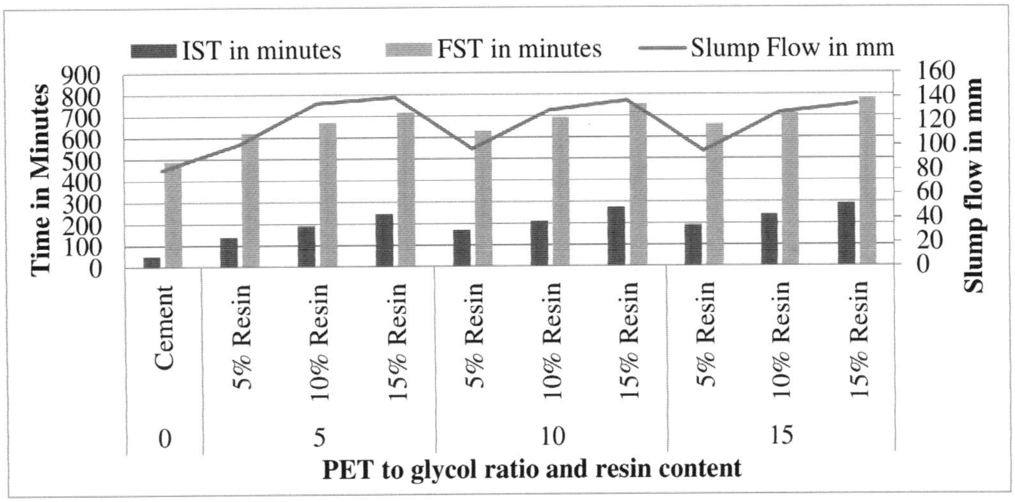

Figure 1. Effect of PET resin content with Various PET to glycol ratios on IST, FST, and Slump flow.

3.2 Soundness of cement
It is essential that cement paste once set should not undergo a large change in volume. No change in the expansion of the cement was found due to the addition of PET resin with various 5%, 10% and 15% PET resin content, as shown in table no 1. It is also notified that there is no effect on soundness for various PET to glycol ratio.

3.3 Flow table test on mortar

It was observed that there is an increase in slump flow of mortar with the addition of various PET resin content, as shown in figure 1. Maximum slump flow was observed for 15% PET resin content for all PET to glycol ratio. Slump flow increases 18.75 % to 26.25%, 58.75% to 68.75%, and 67.5% to 75% for 5%, 10%, and 15% PET resin content for 15%, 10%, and 5% PET to glycol ratio respectively.

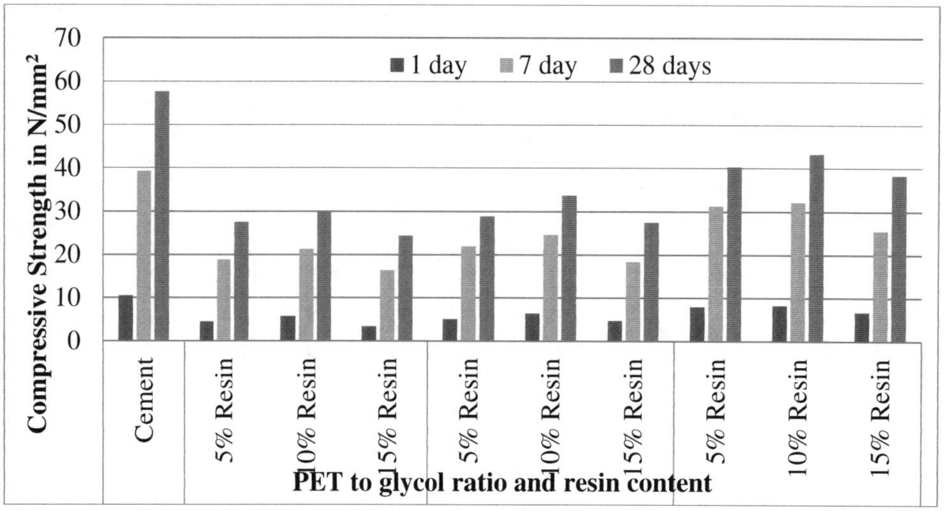

Figure 2. Effect of PET resin content with Various PET to glycol ratios on compressive strength with normal water curing.

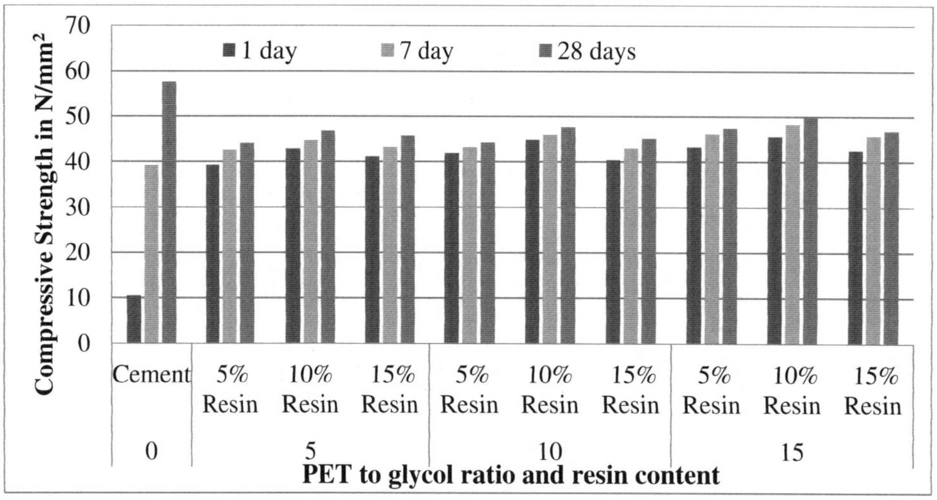

Figure 3. Effect of PET resin content with Various PET to glycol ratios on compressive strength with oven-dry curing.

3.4 Setting time

Initial setting time is increased as a percentage of PET resin content goes on increasing. For 5%, 10%, and 15% PET resin content initial setting time (IST) increases from 90 min to 240 min as compared to control cement paste for 5%, 10% and 15% PET to glycol ratio. The same is shown in figure 1. This could be the advantage for ready-mix concrete. Final setting time (FST) is also increased from 130 minutes to 290 minutes for various PET rein content 5%, 10%, and 15% for PET to glycol ratio of 5 %, 10%, and 15% respectively. Setting time is mainly caused by hydration of C3A and C3S and accompanied by temperature rise in the cement paste initial set correspondence rise of temperature final set correspondence to the peak temperature. Addition of PET resin with a different PET to glycol ratio slows down the temperature rise at initial and peak level due to this, there is increasing in the initial and final setting time.

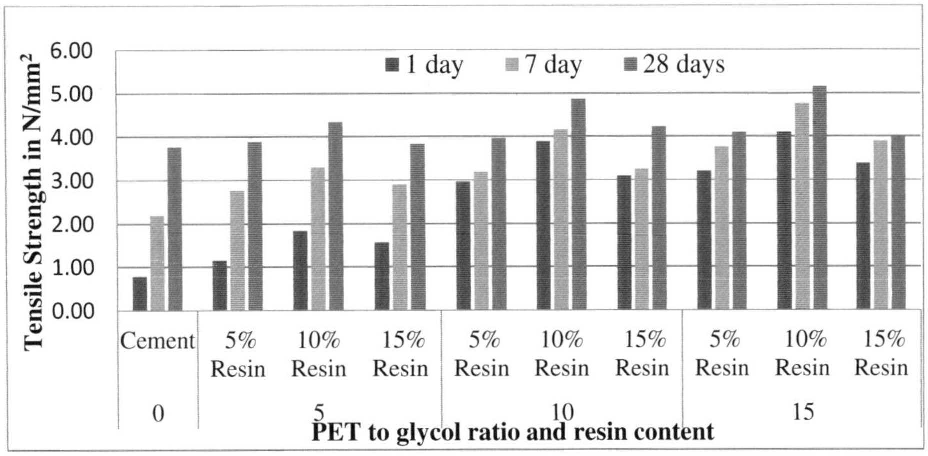

Figure 4. Effect of PET resin content with Various PET to glycol ratios on Tensile strength with oven-dry curing.

3.5 Compressive strength

1, 7, and 28 days compressive strength of the cement mortar was determined in oven dry and normal water curing. In 5% PET to glycol ratio, there is a reduction in compressive strength in comparison with normal cement mortar. After 1-day water curing mixes with 5%, 10%, and 15% PET resin respectively, along with curing agent gained 42.69%, 55.21%, and 31.89% less compressive strength.[16] While compressive strength at the age of 28 days was 47.75%, 51.76%, and 42.24% less in comparison with normal cement mix, as shown in figure 2. However, in oven-dry curing respectively, 274.98%, 309.17%, and 293.03% increase in compressive strength was observed for 1 day curing. However, at 28 days curing the compressive strength reduces respectively by 23.5%, 18.8%, and 20.72% with the addition of PET resin 5 %, 10%, and 15% respectively same is shown in figure 3.

This decrease in the strength could be because of the separation of the interface between aggregate[17] and PET resin and due to the addition of PET resin there might be absence of C3S and C2S most important compound which are responsible for hydration of cement paste

Similar trend is observed with10% and 15% PET to glycol ratio for 5%, 10% and 15% PET resin content. In oven-dry curing, it is observed that cement gains early strength at 1 day curing as compared to normal cement mortar but at 28 days curing strength normal cement mortar has more strength as compared to PET resin substitute. Oven dry curing gives us good strength because of BPo which reacts with PET resin [18] and forms the ammonia curing agent. Among all the PET to glycol ratio, 15% PET to glycol ratio for mixture with 10% of resin content have shown good impact over substitute of cement.

3.6 Tensile strength

Tensile strength of cement mortar was determined at 1, 7 and 28 days in oven-dry curing. There is an increase in tensile strength of mortar for PET resin content and for all PET to glycol ratio as compare to normal cement mortar specimen. In 5% PET to glycol ratio there is increase in tensile strength for 1 day curing 46.67%, 133.33% and 98.78% similarly for 28 days curing 3.26%, 15.21% and 1.67 % as compare to normal cement mortar for 5%, 10% and 15% resin content respectively. A similar trend is observed for 10% and 15% PET to glycol ratio. Among all the PET to glycol ratio and PET resin content 15% PET to glycol ratio and 10%, resin content gives better strength. In all PET resin content, 10% resin gives higher strength for all PET to glycol ratio. The increase in tensile strength is because of the bridging action of the PET particle. The increase in tensile strength is advantageous in rigid pavement work. On use of PET resin in combination with cement there is surface grafting between PET and cement by covalent bond formation between hydropobic PET and hydrophilic alumina silicate of cement.

4 Conclusion

1) With the use of PET resin, a negligible reduction is seen in the amount of water required for standard consistency.
2) There is no effect over the soundness of cement for all PET resin content.
3) Initial setting time, final setting time and slump flow increases as PET resin content increases in comparison with normal cement content specimen.
4) PET resin and BPo shows a reduction in compressive strength at all ages of the normal water curing and gain in compressive strength under oven-dry curing.
5) PET resin show an increase in tensile strength of mortar in oven-dry curing in comparison with normal cement mortar
6) At all PET resin content and PET to Glycol ratio, oven-dry curing shows a positive impact on compressive strength at 1-day curing age in comparison with the normal specimen. However, at 28 days curing age, the compressive strength is less in comparison with a normal mix. This signifies that PET resin can be used in the applications where early strength is required in consideration.

5. References

[1] R. Bedi, R. Chandra, and S. P. Singh, "Mechanical Properties of Polymer Concrete," vol. 2013, 2013.

[2] Z. Yao, X. Zhang, Z. Ge, Z. Jin, J. Han, and X. Pan, "Mix Proportion Design and Mechanical Properties of Recycled PET Concrete," *J. Test. Eval.*, 2015.

[3] C. Marthong and D. K. Sarma, "Influence of PET fiber geometry on the mechanical properties of concrete: An experimental investigation," *Eur. J. Environ. Civ. Eng.*, vol. 20, no. 7, pp. 771–784, 2016.

[4] N. Saikia and J. De Brito, "Use of plastic waste as aggregate in cement mortar and concrete preparation: A review," *Construction and Building Materials*. 2012.

[5] C. Marthong, "Effects of PET fiber arrangement and dimensions on mechanical properties of concrete," *IES J. Part A Civ. Struct. Eng.*, 2015.

[6] B. W. Jo, G. H. Tae, and C. H. Kim, "Uniaxial creep behavior and prediction of recycled-PET polymer concrete," *Constr. Build. Mater.*, 2007.

[7] J. Kim, D. Jeong, C. Son, Y. Lee, E. Kim, and I. Moon, "Synthesis and applications of unsaturated polyester resins based on PET waste," vol. 24, no. 6, pp. 1076–1083, 2007.

[8] L. A. Pereira De Oliveira and J. P. Castro-Gomes, "Physical and mechanical behaviour of recycled PET fibre-reinforced mortar," *Constr. Build. Mater.*, 2011.

[9] L. Agavriloaie, S. Oprea, M. Barbuta, and F. Luca, "Characterisation of polymer concrete with epoxy polyurethane acryl matrix," vol. 37, pp. 190–196, 2012.

[10] K. S. Rebeiz, "Precast use of polymer concrete using unsaturated polyester resin based on recycled PET waste," *Constr. Build. Mater.*, 1996.

[11] E. Rahmani, M. Dehestani, M. H. A. Beygi, H. Allahyari, and I. M. Nikbin, "On the mechanical properties of concrete containing waste PET particles," *Constr. Build. Mater.*, 2013.

[12] F. Mahdi, H. Abbas, and A. A. Khan, "Flexural, shear and bond strength of polymer concrete utilizing recycled resin obtained from post-consumer PET bottles," *Constr. Build. Mater.*, 2013.

[13] Q. F. Yue, L. F. Xiao, M. L. Zhang, and X. F. Bai, "The Glycolysis of Poly(ethylene terephthalate) Waste: Lewis Acidic Ionic Liquids as High Efficient Catalysts," pp. 1258–1271, 2013.

[14] J. Purohit, G. Chawada, B. Choubisa, M. Patel, and B. Dholakiya, "Polyester Polyol Derived From Waste Poly (Ethylene Terephthalate) for Coating Application on Mild Steel Polyester Polyol Derived From Waste Poly (Ethylene Terephthalate) for Coating Application on Mild Steel," vol. 2012, 2012.

IOP Conf. Series: Materials Science and Engineering **829** (2020) 012005 doi:10.1088/1757-899X/829/1/012005

[15] S. Yang, X. Yue, X. Liu, and Y. Tong, "Properties of self-compacting lightweight concrete containing recycled plastic particles," *Constr. Build. Mater.*, 2015.

[16] N. Saikia and J. De Brito, "Mechanical properties and abrasion behaviour of concrete containing shredded PET bottle waste as a partial substitution of natural aggregate," *Constr. Build. Mater.*, 2014.

[17] V. Morin, M. Moevus, I. Dubois-Brugger, and E. Gartner, "Cement and Concrete Research Effect of polymer modification of the paste – aggregate interface on the mechanical properties of concretes," vol. 41, pp. 459–466, 2011.

[18] B. Dä and L. Lichołai, "The effect of the type of curing agent on selected properties of epoxy mortar modified with PET glycolisate," *Constr. Build. Mater.*, vol. 124, pp. 11–19, 2016

IOP Conf. Series: Materials Science and Engineering **829** (2020) 012006 doi:10.1088/1757-899X/829/1/012006

Adaptability of Materials in Green Buildings: Australian Case Studies and Review

Koorosh Gharehbaghi[1,a], **Farshid Rahmani**[1,b] **and David Paterno**[1,c]
[1]RMIT University, Melbourne, Australia

[a] koorosh.gharehbaghi@rmit.edu.au, [b] farshid.rahmani@rmit.edu.au, [c] david.paterno@rmit.edu.au

Abstract. Globally, more and more buildings are constructed as the basis of green methodology. Green building methodology includes the reduction of energy from the conception through the operation phases. Importantly, the practice of green building is gradually becoming more regulatory compliance rather than optional. Subsequently, this paper will carefully review the adaptability of materials in green buildings. In supporting such aim, this paper will also reviewed five Australian case studies. For these building, as a part of adaptability of green materials, recycled concrete, metal and timber were utilized. It was found that most benefits of adaptability of materials for these buildings were significantly improved with the integration with sustainable design. Such alignment is the key for effective material selection and usage in green buildings. The use of green materials alone may be not be enough to considerably reduce the energy usage. Holistically, the green design together with recycled materials need to be combined to deliver efficient green buildings.

1. Introduction

Universally, green buildings refer to advanced and innovative processes which ultimately lead to less environmental impacts [1-4]. Such processes include all the building design and construction together with operational practices. On the other hand, an important aspect of green buildings are the utilized materials [5-7]. Sodagar et al. [8] argued that, the impact of green buildings can at times be heavily material based. Another words, a more sustainable building approach can be undertaken with the limiting material sources and thus the subsequent build-up of waste in landfills. Nonetheless, the green building methodology includes all the mining and collection of materials, their preparation and production, the transportation and the overwhelming creation of construction waste [9]. Additionally, the practice of green building is gradually becoming more regulatory compliance rather than optional [10-12]. Such outcome is based on universal reduction of global resources to high levels of pollution for the benefits of environment [13-15]. Subsequently, all the entities involved in building and construction field such as architects, manufacturers and so on need to carefully consider the environment conservation [16-18]. Subsequently, this paper will carefully review the adaptability of materials in green buildings. In supporting such aim, this paper will also review five Australian case studies.

2. Literature review

Traditionally, due to renewable focus, the green building materials have nontoxic properties which are made from recycled materials [19-22]. Moreover, these can be recycled and subsequently share the characteristics of being both energy and water efficient. For green buildings such materials share distinct benefits: ability to conserve energy; increased adaptability through the design stages and so on [23-25]. Such benefits will ultimately lead to overall reduced costs ie moderate repair costs throughout the building's lifecycle [26-28].

As far as comparison of such materials in opposition to the traditional i.e. steel, Gunn [29] signified the that complex deterioration can be expected in the form of replacement and repair for the standard building materials, however in contrast, for green materials they can be less complex. However, the real advantage of green materials is their overall energy consumption. Green materials such as stone, due to their the natural bases, generally require very little maintenance and rehabilitation. Additionally, stone has far superior water resistance in comparison to concrete. Morel et al. [30] further compared energy consumption of stone opposed to concrete, with their result suggesting that typically a concrete house would consume couple of times more energy. They also noted that transportation of concrete when compared to stone has even more energy consumption. Subsequently, other research have also highlighted stone as a better alternative to concrete, wood and steel. Particularly, when the durability, maintenance, energy usage and environmental impact of such materials are considered.

Nonetheless, Bartlett and Howard [31] deliberated that there is a common misconception with green buildings costing 5% to 15% more to construct when compared to standard buildings. They further expand on their claim by suggesting that cost advisors in the building industry severely over-estimate the central costs of green building and critically under-estimate the possible cost savings. This view is supported by Nalewaik [32] who also claims that green building green benefits from lifecycle cost savings and that some aspects of such design methodology are more cost-efficient. Subsequently, it can be determined that the general conception of green buildings is associated higher costs throughout their life. Gharehbaghi and Rahmani [33] reviewed the concept of green materials, as alignment with recycled substances particularly, concrete, metal and to extent degree timber. While recycled concrete usually consists of crushed materials, it can be used as dry aggregate for many purposes such as new mixes. On the other hand, recycled metal would generally include scraped aluminum, copper, steel, brass, iron and so on. Both recycled concrete and metal can significantly reduce the overall cost of material's cost [3, 34, 35]. At the same time as, recycled timber could also be reprocessed as alternative building material.

Subsequently, reusing and recycling concrete, metal and timber would ultimately reduce the material waste and more importantly lessen environmental impacts [36]. Appropriately, further investigate is required to carefully review both positive and negative possibilities of green buildings and their material usages.

3. Research methodology

This research first reviews the most relevant and recent green building and material literature. In supporting such aim, this paper will also review five Australian case studies. The appropriate data was gathered from used various building log-books together with design schematics and reports were accessed and scrutinized. Finally, since this study is a qualitative investigation, an overall finding was then evaluation

4. Case studies

The five buildings were all randomly selected, from different Australian estates and were based on two classifications - retrofitting and new buildings. The first three homes were built with standard building design and then retrofitted with green building initiatives. The remaining two homes were built as brand new and utilized green building design. Table 1 summarizes the five case studies.

IOP Conf. Series: Materials Science and Engineering **829** (2020) 012006 doi:10.1088/1757-899X/829/1/012006

Table 1. Overview of the case studies.

Building location	Main green building features	Main green materials
Building 1: Birkenhead, South Australia. *Retrofitting*	The renovation of a single-story home that focused on making the structure energy efficient with limited area and spending. Its improvements were: Increasing thermal mass, improving windows and glazing, SIPs wall construction, sealed building fabric, solar hot water and solar photovoltaic systems, energy-efficient lighting and water saving. Each improvement is detailed in the section of the case study and identifies how each of the improvements can be incorporated into future construction builds.	Recycled concrete, metal and timber
Building 2: Curtin, Australian Capital Territory. *Retrofitting*	The renovation of a two-story home that focused on improving temperature control during colder periods and increasing the availability of natural lighting. Its improvements were based on: Solar hot water and solar photovoltaic systems, passive heating and cooling, active heating and cooling, green building materials and insulation, energy-efficient lighting, recyclable materials and water saving. Each improvement is detailed in the section of the case study and identifies how each of the improvements can be incorporated into future construction builds.	Recycled concrete and timber
Building 3: Fitzroy North, Victoria. *Retrofitting*	The renovation of a two-story home that utilised the following green building initiatives: Passive heating and cooling, sustainable material use, improved windows and glazing features, water saving, energy-efficient lighting, energy saving, solar hot water and solar photovoltaic systems and insulation. Each improvement is detailed in the section of the case study and identifies how each of the improvements can be incorporated into future construction builds.	Recycled concrete, metal and timber
Building 4: Caloundra, Queensland. *New building*	The development of a new display home that focused on displaying the potential of construction with green building initiatives. Its inclusions were: Reverse brick veneer, lightweight material construction, insulation, glazed windows, solar photovoltaic system, energy-efficient appliances and water saving. Each inclusion is detailed in the section of the case study and identifies how each of the inclusions can be incorporated into future construction builds.	Recycled concrete, metal and timber
Building 5: Darwin River, Northern Territory. *New*	The development of a two-story home that focused on passive cooling and water saving. Its green building inclusions were: Water saving, embodied energy reductions, greenhouse gas reductions and renewable sources of energy. Each inclusion is	Recycled concrete, metal and timber

IOP Conf. Series: Materials Science and Engineering **829** (2020) 012006 doi:10.1088/1757-899X/829/1/012006

building	detailed in the section of the case study and identifies how each of the inclusion can be incorporated into future construction builds.	

Importantly, all the five buildings utilized range of cooling and heating initiatives including, solar hot water systems, solar photovoltaic systems. These initiatives were together used with green materials as a part of the renewable energy production and greenhouse gas reduction strategies. As a part of such materials selection recycled concrete, steel and timber were utilized. As a concept of adaptability of materials in green buildings, this arrangement included lightweight construction with utilization of excess materials from a different demolition while using recycled timber. Foundation slab: 60% recycled materials and 30% cement substitute. Removal of bricks at various sections of the structure and replaced with restored timber weatherboards in the form of cladding. Change to the existing walls insulation that included R3 recycled polystyrene, R2 polyester batts and R5 wool cells. For the existing homes, materials used generally that had lower impact on the environment while having reductions in embodied energy. In addition, the renovation included facade alterations including change to the existing walls insulation that included R3 recycled polystyrene, R2 polyester batts and R5 wool cells. These changes restricted the flow of heat during winter to maintain warmth within the home, whereas the old design lacked the insulation required to do so. Generally, the use of the green processes and materials saved the buildings over a tonne of CO_2.
Further, all the five buildings utilized three main green materials, recycled concrete, steel and timber. The overall percentile of materials in the five green buildings in presented in figure 1.

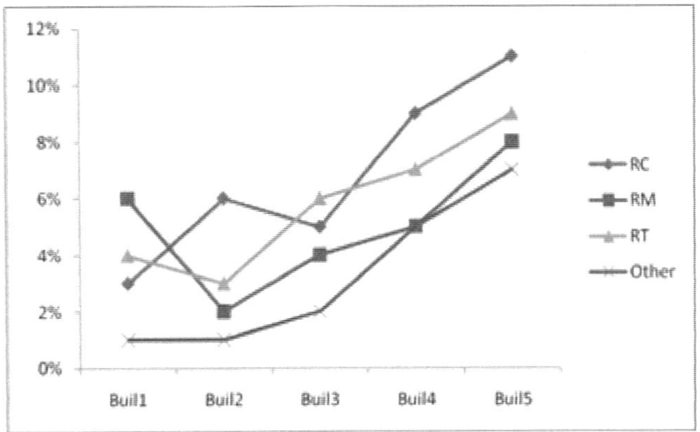

Figure 1. Comparison of green materials for five buildings.

The adaptability of the "overall" materials in the five green buildings in represented in Figure 1. For these five buildings, it seems that recycled concrete is the most dominate green material. The reason for such usage, could be due to more alternatives to use recycled concrete. Particularly for pre-case panels together with low-load bearing walls recycled concrete is an ideal material. Nonetheless, both recycled metal and timber were comparably used in these five buildings. However, the adaptability of materials in these structures were also further supported by green building design. Another words, to reduce energy usage, the materials in these buildings were cooperated with green design. Such green design included, improving the natural lighting through better orientation and new design windows.
Moreover, as a part of energy self-efficiency, these buildings sealed off penetrations to greatly improve passive heating and cooling. These buildings use such strategies to provide a passively heat and cool which in-turn reduce the amount of necessary active heating and cooling. The use of green materials along with other energy reductions plans such as rainwater tanks significantly reduced the energy usage for these buildings. The lesson which can be learned, is that the use of green materials alone may be not

enough to considerably reduce the energy usage. The green design together with recycled materials go together to deliver efficient green buildings.

5. Conclusion

This paper aimed to review the adaptability of materials in green buildings, to improve the overall understand of green building construction methods and materials. Subsequently, other considerations such as the general benefits of using standard green building materials and methods were also researched. To further strengthen the research aim, five Australian case studies were identified and analyzed. Aptly, each study demonstrated how the different regions and climates can ultimately adapt affect the structures and their materials. Nonetheless, as the basis of adaptability of materials in green buildings the findings of the research can be summarized as; a) generally, there are several small changes which can be made to produce a structure which is the home environmentally friendly and energy exertion, and b) such initiatives can be significantly cost efficient. As the findings of the case studies showed, most of the cost effectiveness does not only occur with the material selection but rather the mutually of sustainable design to support the green building adoptability.

6. References

[1] R. Riles, M. Bilec, N. Gokhan, and K. Needy: The Economic Benefits of Green Buildings: A Comprehensive Case Study, The Engineering Economist, Volume 51, Issue 3 (2006), p. 259-295.

[2] U. Berardi, and A. Ghaffarian: State-of-the-art analysis of the environmental benefits of green roofs, Applied Energy, Volume 115 (2014), p. 411-428.

[3] K. Gharehbaghi, and C. Scott-Young: Waste diminution in construction projects: Environmental predicaments , Earth and Environmental Science, Institute of Physics, United Kingdom, vol. 127, no. 1 (2018), pp. 58-63.

[4] A. Arıoğlu, Ö. Mahmure, D. Dhavale, and J. Sarkis: Greenhouse gas emissions in the construction industry: An analysis and evaluation of a concrete supply chain, Journal of Cleaner Production, vol.167 (2017), pp.1195-1207.

[5] B. Reed, R. and Fedrizzi: The Integrative Design Guide to Green Building Redefining the Practice of Sustainability, Hoboken, Wiley, (2011).

[6] K. Gharehbaghi, and M. Georgy: Sustainable Construction by Means of Improved Material Selection Process, International journal on the Academic Research Community publication, Volume 1, issue 1 (2019), pp 85-94.

[7] J. Giesekam, J. Barrett, and P. Talyor: Construction sector views on low carbon building materials, Building Research & Information, vol 44 (2015), pp. 423-444.

[8] B. Sodagar, D. Rai, B. Jones, J. Wihan, and R. Fieldson: The carbon-reduction potential of straw-bale housing , Building Research & Information, Volume 39, Issue 1 (2010), p. 51-65.

[9] F. Ching, and M. Shapiro: Green building illustrated", Hoboken, New Jersey, Wiley (2014)

[10] P. Melia, G. Ruggieri, S. Sabbadini, and G. Dotelli: Environmental impacts of natural conventional building materials: a case study on earth plasters, Journal of Cleaner Production, Volume 80 (2014), p. 179-186.

[11] O. Akadiri: Understanding barriers affecting the selection of sustainable materials in building projects, Journal of Building Engineering, vol. 4 (2015), pp. 86-93, Elsevier.

[12] B. Guerra, A. Bakchan, F. Leite, and K. Faust: BIM-based automated construction waste estimation algorithms: The case of concrete and drywall waste streams, Waste Management, Vol. 87 (2019), pp.825-832.

[13] G. Ding: Sustainable construction – The role of environmental assessment tools, Journal of Environmental Management, Volume 86, Issue 3 (2008), p. 451-464

[14] J. Glasson, R. Therivel, and A. Chadwick: Introduction to Environmental Impact Assessment", Taylor and Francis, UK (2012), pp 21-43.

[15] B. Hadavand, and R. Imaninasab: Assessing the influence of construction and demolition waste materials on workability and mechanical properties of concrete using statistical analysis, Innovative Infrastructure Solutions, vol. 4, no. 1. (2019).

[16] R. Spiegel, and D. Meadows: Green building materials: A guide to product selection and specification, Hoboken, New Jersey, Wiley, (2012).
[17] C. Kilbert: Sustainable construction: Green building design and delivery, Hoboken, Wiley (2016), pp 9-14.
[18] C. Jones: Wooden buildings reach new heights' Materials World magazine, Vol 270219 (2019).
[19] K. Gharehbaghi: Advancements in concrete technology in Australia: geo-polymer concrete, International Journal of Constructed Environment, Volume 7, issue 1 (2015), pp 19-29.
[20] E. Fehling, B. Middendorf, and J. Thiemicke: Ultra-High Performance Concrete and High Performance Construction Materials, Kassel University Press, Kassel, Germany (2016), pp 59-67.
[21] V. Kashintseva, W. Strielkowski, J. Streimikis, and T. Veynbender: Consumer attitudes towards industrial CO2 capture and storage products and technologies, Energies, vol. 11, no. 10 (2018).
[22] A. Lukman, O. AkanbiacLukumon, O, Olugbenga Akinadea, A. AjayiaJuan, M. Davila, D. Hakeem, and A. Owolabia: Disassembly and deconstruction analytics system (D-DAS) for construction in a circular economy, Journal of Cleaner Production, Vol. 223 (2019), pp. 386-396.
[23] J. Allwood: Sustainable materials – with both eyes open, Cambridge, University of Cambridge (2012).
[24] K. Gharehbaghi, and R. Chenery: Fiber Reinforced Concrete (FRC) for High Rise Construction: Case Studies, Materials Science and Engineering, vol. 272 (2017), pp 223-229.
[25] S. Lockrey, K. Verghese, E. Crossin, and H. Nguyen: Concrete recycling life cycle flows and performance from construction and demolition waste in Hanoi, Journal of Cleaner Production, vol. 179 (2018), pp. 593-604.
[26] E. Simiu, and D. Yeo: Advances in the design of high-rise structures by the wind tunnel procedure: Conceptual framework, Wind and Structures, An International Journal, vol. 21, no. 5 (2015), pp. 489-503.
[27] W. Salama: Design of concrete buildings for disassembly: An explorative review, International Journal of Sustainable Built Environment, vol. 6, no. 2 (2017), pp. 617-635.
[28] K. Reddy, C. Cameselle, and J. Adams: Engineering Practices, Sustainable Engineering: Drivers, Metrics, Tools, Engineering Practices, and Applications, Vol. 1, no.1 (2019), pp 419.
[29] S. Gunn: Stone House Construction, Australia, CSIRO Publishing, (2012).
[30] C. Morel, A. Mesbah, M. Oggero, and P. Walker: Building houses with local materials: means to drastically reduce the environmental impact of construction, Building and Environment, Volume 36, Issue 10 (2001), p. 1119-1126.
[31] E. Bartlett, and N. Howard: Informing the decision makers on the cost and value of green building, Building Research & Information, Volume 28 (2008), p. 315-324
[32] A. Nalewaik: Costs and Benefits of Building Green, American Society of Engineers: Proceedings of the 5th International Engineering and Construction Conference, August 27-29, Los Angeles, U.S.A, p. 969-979 (2008).
[33] K. Gharehbaghi, K., and F. Rahmani: Practicalities and Developments of High-Rise Composite Structures: Case Studies, Materials and Technologies in Engineering (2018), pp 153-159.
[34] K. Gharehbaghi and C. Scott-Young: GIS as a vital tool for Environmental Impact Assessment and Mitigation", Earth and Environmental Science, Institute of Physics, United Kingdom, vol. 127, no. 1. (2018a), pp. 52-57.
[35] K. Gharehbaghi, M. Georgy, M., F. Rahmani: Composite High-Rise Structures: Structural Health Monitoring (SHM) and Case Studies, Materials and Technologies in Engineering, (2018). pp 146-152.
[36] M. Pritchard and A. Pitts: Evaluation of Strawbale Building: Benefits and Risks, Architectural Science Review, Volume 49, Issue 4, (2006). pp. 372-384.

IOP Conf. Series: Materials Science and Engineering **829** (2020) 012007 doi:10.1088/1757-899X/829/1/012007

A Study of the properties and microstructure of high-magnesium nickel slag powder used as a cement supplement

L Oksri-Nelfia[13], Reynaldi Akbar[1], and Sotya Astutiningsih[2]

[1]Departement of Civil Engineering, Faculty of Civil Engineering and Planning, Trisakti University, Jakarta 11440, Indonesia
[2]Departement of Metallurgical and materials Engineering, University Indonesia, Depok 16424, Indonesia

[3]E-mail: lisa@trisakti.ac.id

Abstract. This research aims to use high-magnesium nickel slag crushed into a powder as a replacement for cement and to study their properties and microstructure in concrete. The particle diameter of the nickel slag powder was obtained lower than 75 μm after a process of crushing and sieving. The specific areas, absolute dry densities and particle size distribution (PSD) were analysed to obtain the physical properties of the material. The techniques applied to study their chemical and mineralogical composition were the X-Ray Florescence (XRF), the Scanning Electron Micrograph (SEM). The OPC type 1 was used as a reference in comparison to the experimental material of this research. Without a chemical admixture, the compressive strength of concrete after 28 days is fc' 30 MPa and the specific surface area is 4360 cm^2/gr. The compressive strength of high-magnesium nickel slag powder is capable of replacing cement by 20% without altering properties of concrete. According to the XRF technique results, the nickel slag contains silica, calcium, magnesia, and alumina. Furthermore, the image of the SEM-EDS illustrated the slag particle to be sharp, irregular, and containing calcium, silicon, alumina, oxygen, and magnesium. This result confirmed that the XRF technique and SEM-EDS were accurate.

1. Introduction

Indonesia is a country that has many natural resources with numerous mining activities including nickel, iron, gold, coal. However, it is also the third source of environmental pollution in the country, so an environmental consciousness for mining expansion needs to be developed. This must start at the beginning of production and during the process, to enable a sustained establishment [1]. In several mining sites, there is slag waste and there are several cases of B3 pollution (harmful and toxic substance) released from mining sites and so far, no solution has been provided [2]. Currently, the nickel mining produces a redundant nickel slag and its fusion in the domestic sector ranging up to 5 million tons/year with the assumption of producing NPI or FeNi (Ni level is 10%). The standard input of Ni ore substance is approximately 40 million tons/year, which means that 30 million tons become slag [3].

Therefore, this research aims to explore the use of nickel slag and to prevent its effect on the environment. A nickel slag is a solid waste product that is obtained from the ore quarry process [4]. Therefore, the waste system needs to be organized properly to prevent environmental problems. Research has been undertaken to find solutions using the powdered slag obtained from nickel to minimize environmental problems. The compressive strength of the concrete has been improved by replacing sourced materials for concrete with waste the nickel slag aggregates [5]. According to the recent research of using green construction (eco-friendly), the cement quantity can be modified to pozzolan substance such as fly ash [6]. The purpose of this type of concrete is to reduce environmental problems such as global warming, the overexploitation of natural resources, and the waste obtained from the quarries, households, or construction sites. A replacement substance for cement needs to be distinguished as a substitute material, so it can be used in the concrete production.

The use of cement in building construction or other infrastructure work creates negative impacts on the environment. The literature [7] revealed that each ton of cement production generates nearly 0.8 tons of carbon dioxide, which is released into the atmosphere. It is caused from using a large use of energy and a high temperature of approximately 1450°C to produce the clinker, which is the main component of cement production [8, 9]. Hence, the industrial cement causes 5% of the emission from the greenhouse gases [10, 11, 12]. In some research, the cement was partially replaced by pozzolanic wastes product known as fly ash, which comes from coal-burning electricity generation, and ground granulated blast furnace slag from iron production waste, also used are gypsum, Limestone filler. Such methods will reduce the carbon dioxide emission that comes from its clinker production. It is used to produce the eco-friendly concrete by utilizing the industrial production residue (Green Concrete). The substitution of cement is effective for reducing the carbon dioxide emission during production, especially when it is combined with an additional substance [12, 13]. The present research [14, 15], the C-A-S ternary diagram approach is a reaction of the nickel slag powder with water. When it is mixed with water it becomes portlandite reaction $Ca(OH)_2$.

The objective of this research is to use high magnesium nickel slag powder as partial substitute for cement, with the maximum substitution rate up to 30%. The nickel slag powder undergoes a test to determine its physical, chemical, and mineralogical characteristics. The research illustrates the following tests: the pycnometer, the Blaine test, workability, compressive strength and the new concrete production made with a mixture of the nickel slag powder. The compressive strength were conducted in a time phase of 2, 7, 14 and 28 days, with substantial replacement composition of 5%, 10%, 20% and 25% to the weight of the cement. Concrete samples with different substitutions, were named successively NSP5, NSP10, NSP15, NSP20 and NSP25. The concrete with NSP samples was compared to concrete made with 100% OPC type 1 as a reference concrete.

2. Materials and methods

2.1. Preparation of high-magnesium nickel slag powder (NSP)

The nickel slag used was obtained from Southeast Sulawesi, Indonesia. This raw material was prepared by using industrial ball mill came from PT. Growth Java, Cilegon. Indonesia. In order to obtain a granular size less than 0.075 mm, the NSP (nickel slag powder) was filtered by sieving it with a machine strainer of #200 in a dry condition (room temperature between 20-25 °C). After the sieving and straining process that took 10 minutes, the NSP was produced (Figure 1). In addition, to determine the absolute dry density, the nickel slag should be in a dry condition; hence, it was kept in an oven for ± 24 hours with a temperature of 110 °C and it was later placed in desiccator's for ± 24 hours to ensure the nickel slag was in room temperature condition. After that process, the nickel slag can be used in the next test process.

Figure 1. Sample of NSP.

2.2. *Physical, chemical and mineralogical characterization*

The density of NSP was determined by using a pycnometer according to European standard EN 1097-6 [16]. Specific surface area was determined by Blaine test according to European standard EN 196-6 [17]. The particle size distribution and the median diameter (D_{50}) were measured by laser diffraction in dry suspension. The results of the physical characterization are shown in Table 1. Then, the chemical composition of OPC type 1 and the NSP were obtained from the X-Ray Florescence (XRF) by using EPSILON 5 analyzer from Panalytical, The result is presented in Table 2. According to these results, the main chemical components of nickel slag powder are silica, alumina, calcium, and magnesia. This results are confirmed by literature review [18, 19].

Table 1. Physical analysis of materials studied [20].

Material	Density (g/cm^3)	Blaine (cm^2/g)	D_{50} (μm)
NSP	2.75	4360	15.4
OPC Type 1 Tiga Roda	3.15	3670	14.90

Table 2. Chemical analysis of materials studied.

Chemical composition	OPC Type 1 Tiga Roda	NSP
CaO	63.2	23.58
Al_2O_3	4.96	8.85
SiO_2	18.45	42.03
Fe_2O_3	2.86	1.59
SO_3	2.18	-
MgO	3.52	19.51
K_2O	0.31	0.12
Na_2O	0.15	0.08
LOI	3.42	2.59
$CaCO_3$	-	-
C_3S	68.72	-
C_3A	8.30	-
C_4AF	8.70	-

Figure 2. The microstructure of NSP.

Scanning electron microscope (SEM) characterization was carried out with a Brücker-Zeiss EVO MA10 instrument in order to study the morphology of slag particles. The microstructures of nickel slag powder are shown in Figure 2. 28 day concrete samples (NSP 20) were crushed with a compressive strength machine. A fracture-shaped sample with a maximum thickness of 2.5 cm were tested by SEM. Chemical characterization were determined by testing heavy metal leachability (also called trace metal analysis with EPSILON 4, Malvern Panalytical instrumentation). Leaching tests consisted of subjecting a material to the action of a solvent in order to quantify chemical elements (pollutants or not) released by the material. The aqueous solution was placed with a liquid/solid ratio (water/powder) of 10 L/kg, according to the standard NF EN 12457-2 [21]. The leaching agent was demineralized water and the contact process was carried out over a period of 24 hours. At the end of this period, the liquid/solid separation was carried out by filtration. The purpose of these tests is to allow the quantification of heavy metals, arsenic (As), barium (Ba), cadmium (Cd), chromium (Cr), copper (Cu), nickel (Ni), lead (Pb), selenium (Se) and zinc (Zn) in the studied samples. Most metals are dangerous to the environment beyond a certain threshold. The samples were prepared according to the leaching method.

2.3. Composition of concrete with NSP and OPC type 1

To use the nickel slag as the supplementary cementitious materials (SCMs), the concrete was designed on the Japanese International Standard (JIS A 5004), with almost 56% sand in the total aggregate (volume of fraction of sand fixed) (S/A) and a cement water factor (W/C) of 0.47. For testing fine aggregate and coarse aggregate based on JIS A 5004, the fine aggregate passes the cleanliness test, (Sludge Level ≤ 7.0%). Fine aggregate with specific gravity (SSD) was 2580 kg/m^3 and has an absorption of 0.2%, coarse aggregate with specific gravity (SSD) was 2650 kg/m^3, absorption of 0.23%. Ordinary Portland cement (OPC) Type 1 used was Tiga Roda (Indocement Ltd). Composition of the concrete mixture was compared with the compressive strength target, $f_{c'}$ = 30 MPa). The concrete was made by gradually substituting the mass of cement by the nickel slag powder (NSP) from 0% to 25%. The total samples were 72 specimens. The comparison of water ratio to the cement mixture and NSP (water to binder ratio of mortars) (W/ (A/C) is constant and the water and pasta volume were constant. The details can be seen in Table 3. Compressive strength was tested at the ages 3, 7, 14, and 28 days.

Table 3. Mix proportions (in kg/m^3) and properties of the studied concrete.

	Ref	NSP				
		5	10	15	20	25
OPC type 1 Tiga Roda	400	380	360	340	320	300
Nickel slag powder (NSP)	0	20	40	60	80	100
Gravel 10/20	1030	1030	1030	1030	1030	1030
Sand 0/5	662	662	662	662	662	662
Water	195	195	195	195	195	195
W/C	0.47	0.47	0.47	0.47	0.47	0.47
W/(C+A)	0.47	0.47	0.47	0.47	0.47	0.47
Slump (cm)	2 ± 0	2 ± 0.6	4 ± 0.8	4 ± 0.8	6 ± 1	6 ± 1
28-day compressive strength (MPa)	41.7	39.8	36.1	34.2	30.2	28.4

2.4. Properties of concrete

The research analyses the workability by using the slump test with Abrams cone (H = 150 mm, Ø = 100 mm). Furthermore, a test was conducted on the workability of concrete to determine the value of slumps from each concrete produced in a cylinder form (Ø=100mm and H=200mm). The compressive strength was analysed based on the ASTM C470; within a period of 3, 7, 14 and 28 days with digital compression machine PILOT Press (capacity 2000 kN). Before testing it, the concrete should be cured in a water temperature of 20 °C. In making concrete samples, it should be underlined that concrete was made without the addition of superplasticizer. Indeed, it is known that chemical mixing can modify the hydration kinetics [22].

3. Results and discussion

3.1. *Properties of NSP*

3.1.1. *Physical properties*

The result obtained from the absolute dry density, specific surface area and median diameters are shown in Table 1. The absolute dry density of nickel slag powder was 2.75 t/m^3 and cement was 3.15 t/m^3. The absolute dry density NSP was smaller than the density of cement. In addition, the NSP specific surface area was higher than cement. Therefore, the nickel slag can be a supplementary cementitious materials (SCMs). It also influences the compressive strength and workability result [23]. Figure 3(a) shows that the particle size distribution (PSD) with median diameter (D_{50}) of nickel slag powder (NSP) at 15.4 μm was almost the same as the median diameter of OPC type 1 at 14.90 μm in figure 3(b).

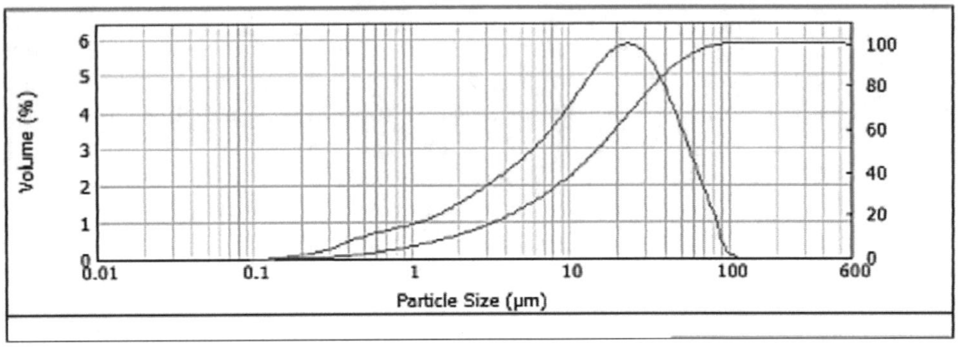

(a)

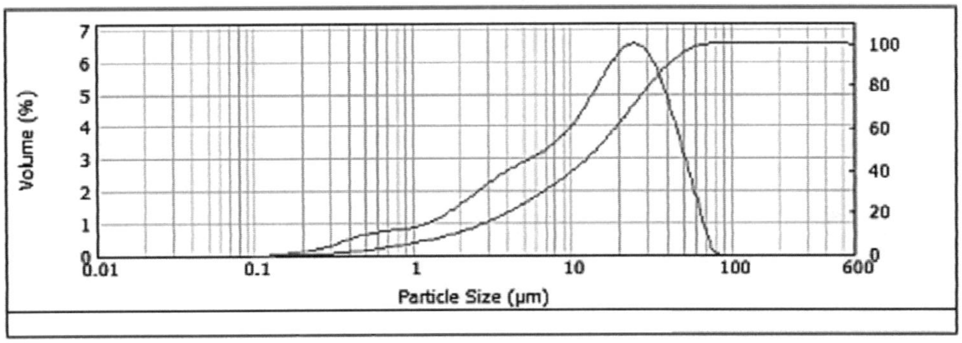

(b)

Figure 3. Particle size distribution of (a) NSP (b) OPC type 1.

3.1.2. *Chemical and mineralogical composition*

According to trace metals testing for raw material analysis, the assay showed that the arsenic, cadmium, and mercury level are below the threshold of nickel slag powder (NSP) as shown in literature [20]. Hence, this substance is very safe for the environment and can be implemented as the new substitute of cement [4]. Cement industry (PT Indocement Ltd) has been accepted criteria for trace element solid waste in this case for this sample.

The figure 4 shows the results obtained from the SEM images and EDS test on the mineral composition for NSP 20. The EDS data results was carried out to identify dominant element of NSP. In figure 4(a) the SEM image was magnified 500 times. Furthermore, it uses a 20 kV SE detector as the age image obtained from a concrete sample for NSP25 at 28 days. From the SEM, there was a 20μm particle size of NSP; the image forms a particle that contains each mineral element, which was explained in the EDS result in figure 4(c), Magnesia, Alumina, calcia, Silica, iron and Oxygen dominate the EDS data result [19]. In addition, figure 4(b) shows that the SEM images were magnified 2000 times (microscope scale magnification) with a 20 kV SE detector. This Figure shows that Silica and Oxygen dominate the EDS results. The colour indicated that it contains dominant Iron (Fe), and it is lighter than the Silica (Si) [24].

(a)

(b)

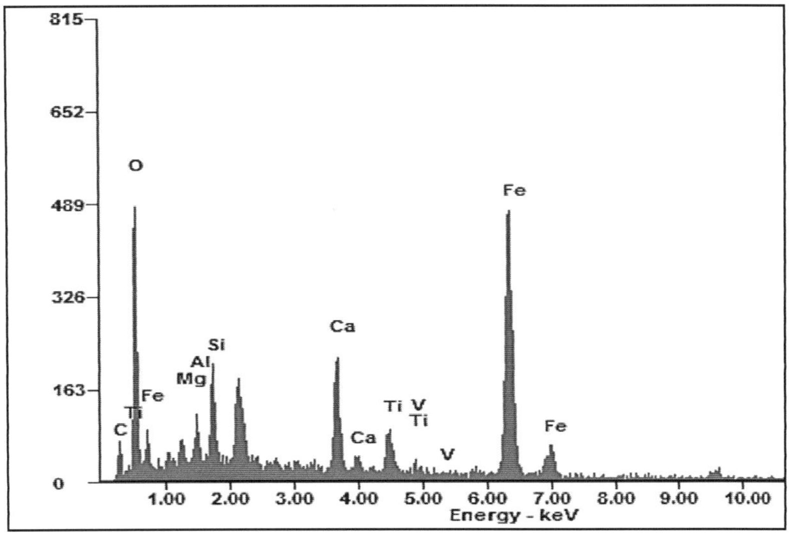

(c)

Figure 4. SEM images NSP20 concrete sample at 28 days in (a) with 500 x of magnification and (b) with 2000x magnification (c) EDS analysis.

3.2. Properties of concrete prepared with NSP

The influence of NSP on concrete consistency is a higher result obtained in the workability test on NSP, whatever the substitution rate, than OPC Type 1 as shown in Table 3. According to literature [25], the workability result in each substitution level is always higher than the material of reference, it means that nickel slag does not absorb water (low water absorptions). This result is surprising considering the fineness of NSP particles finer than cement. In fresh state, recycled material tend to affect the rheology of concretes because it absorbs less water and increases the compactness of the granular skeleton, because of their nature and shape.

The influence of NSP on the compressive strength is based on the data provided in figure 5, the research concludes that the compressive strength for the partial cement substitution is 20% of the cement mass at $f_{c'}$ 30 MPa at 28 days. Above 25% substitution, the compressive strength is below 30 MPa and so on, the compressive strength of the design target. Based on the research [23], it stated that there was a power improvement in each age due to its high-density level and low porosity. It is influenced by a higher nickel refinement. The higher the cement substitution of the nickel slag is, the lower the compressive strength is obtained with different curing times. Based on European standard EN 206-1, the substitution of cement level must not exceed 25% because it will affect the mechanical characteristics and durability of the concrete.

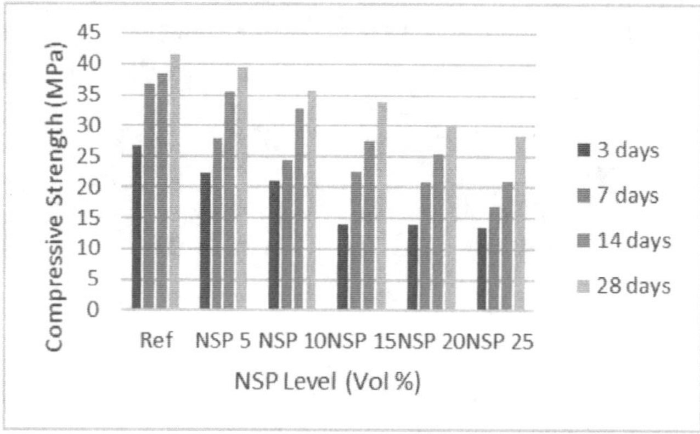

Figure 5. Effect of NSP on compressive strength of concrete.

4. Conclusions

The results from the study showed that NSP can be used as substitution for OPC type 1 in concrete mixture. It is expected to become an alternative construction material. The NSP with a size of 0.075 mm has finer properties and lower water absorptions rate compared to OPC type 1. In general, the obtained affirmative results revealed that NSP can replace up to 20% of the cement in the concrete mixture without altering properties of concrete and it has compressive strength of 30 MPa at 28 days of the concrete. This nickel slag contains a high magnesium level, where the results are proven by XRF and SEM-EDS. However, we need to do further tests of the compressive strength at age of 56, 90 and 180 days to see the pozzolanic activity of this material. It causes a higher compressive strength than normal concrete (without NSP substitution) long term conditions.

5. References

[1] Zainul R, Oktavia B, Dewata I, Efendi J 2018 Study of Internal Morphology on Preparation of Cu_2O Thin-Plate using Thermal Oxidation. *Int Conf on Science and Technology*

[2] Arino A M & Mobasher B 1999 Effect of ground copper slag on strength and toughness of cementitious mixes. *ACI Materials Journal,* **96**(1), 68-73

[3] KEPEMENPERIN 2018 Support of Slag Utilization Policy for Downstream Industry and Infrastructure Sectors

[4] Salain, I, Made A, Sudarsana, K, and Mustika W 2015 Mechanical properties of concrete using nickel slag as coarse mechanical properties of concrete, *International Journal of Engineering Inventions*, Vol 4, (9), p 7-16

[5] Achmadi A L I 2009 Slag Sebagai Agregat Halus Dan Agregat kasar Program Pascasarjana (*Slag as a Fine Aggregate and Coarse Aggregate*) post graduate program report

[6] Azeez Bahedh M, Saleh Jaafar M 2018 Ultra High-Performance Concrete Utilizing Fly Ash as Cement Replacement under Autoclaving Technique, Case studies in construction materials journal, vol 9

[7] Bastier, R 2000 Fours de cimenterie, ateliers de cuisson du clinker. Tech. L'ingénieur

[8] Mehta PK 1999 *Concrete technology for sustainable development*. Concr Int 47–52

[9] World business council for sustainable development (WBCSD) 2009 Cement industry energy and CO performance getting the Numbers Right (GNR)

[10] Humphreys K and Mahasenan M 2002 Toward a sustainable cement, substudy 8: Climate Change, World Business Council for Sustainable Development, Cement Sustainability Initiative

[11] Dalgleish T, Williams J M G, Golden A M J, Perkins N, Barrett L F, Barnard P J, Yeung C, Murphy V, Elward R, Tchanturia K, Watkins E 2007 Handbook of Low Carbon Concrete, Elsevier L. ed, *J. of Experimental Psychology: General. Joe Hayton*

[12] Hendriks A, Worrell D, Jager D, Block K and Riemer P 2000 Emission reduction of greenhouse from the cement industry. *Tech report*, IEA greenhouse gas R&D Program

[13] Flower D J M and Sanjayan J G 2007 Greenhouse gas emissions due to concrete manufacture. Int. J. Life Cycle Assess. **12**, 282–288

[14] Ashad H 2008 Optimisasi Bubuk Slag Nikel dengan Sistem Ternary C-A-S (*Optimization of Slag Powder Nickel with Ternary Systems C-A-S*) Vol 15, No 3.

[15] CIMBéton 2004 La durabilité des bétons. T48. Bétons ouvrages d'art 93

[16] NF EN 1097-6 2014 Essai pour déterminer les caractéristiques mécaniques et physiques des granulats - partie 6 : Détermination de la masse volumique réelle et du coefficient d'absorption d'eau

[17] EN 196-6 1990 Method of testing cement. Part 6: Determination of fineness

[18] Astutiningsih S and Suharno B 2019 NI SLAG VALORISATION IN INDONESIA : A CASE STUDY Slag characterization, *Int Slag Valorisation Symp* (Mechelen, Belgium)

[19] Wu Q, Wu Y, Tong W and Ma H 2018 Utilization of nickel slag as raw material in the production of Portland cement for road construction. *Constr Building Materials* 193, 426–434

[20] Oksri-Nelfia L, Akbar R, Astutiningsih S 2019, KoNTekS 13 Proc, Aceh, Indonesia (To be published)

[21] NF EN 12457-2 2002 Lixiviation - essai de conformité pour lixiviation des déchets fragmentés et des boues

[22] Siler P, Kolarova I, Kratky J. Havlica J and Brandstetr J 2014 Influence of superplastisticizers on the course of hydration of Portland cement hydration Chem.pap.**68**, 90-97

[23] Sugiri, S 2005 Penggunaan Terak Nikel sebagai Agragat dan Campuran Semen untuk Beton Mutu Tinggi (Use of Nickel Slag as an Aggregate and Cement Mixture for High Quality Concrete), *J. infrastructure and Built Environment*, Vol 1 No 1

[24] Sujiono E H, Zharvan V, Yunita S N and Samnur S 2018 Influence of PT Vale's Granulated Gradation Nickel Slag Aggregate on Compressive Strength of Concrete *IOP Conf. Series: Materials Science and Engineering*

[25] Oksri-Nelfia L, Mahieux P Y, Amiri O, Turcry P, Lux, J 2016 Reuse of recycled crushed concrete fines as mineral addition in cementitious materials. *Mater. Struct. Constr.* **49**, 3239–3251

Acknowledgements

The author would like to acknowledge Budi Hartono (PT Indocement Ltd, Citereup) for chemical analyses by heavy metal leachability and PT Jaya Beton Indonesia (JBI) for testing compressive strength and also PT Growth Java, Cilegon Indonesia to provide nickel slag powder.

IOP Conf. Series: Materials Science and Engineering **829** (2020) 012008 doi:10.1088/1757-899X/829/1/012008

A Study on Properties of Concrete Made with Processed Granulated Blast Furnace Slag as Fine Aggregate

Venkataramu Vibha[1,a] and B V Venkatarama Reddy[2,3]

[1,2]Department of Centre for Sustainable Technologies, Indian Institute of Science Bangalore, India,
[3]Department of Civil Engineering, Indian Institute of Science Bangalore, India

[a] vibhav@iisc.ac.in

Abstract. This study focusses on understanding the influence of processed granulated blast furnace (PGBS) slag on fresh and hardened state properties of concrete. Nowadays, metallurgical slags are being explored as alternative materials to natural virgin aggregates in concrete. Before using any such material in concrete, it is important to understand its influence on properties of concrete. In this study an attempt has been made to use granulated blast furnace slag from a local steel plant as a substitute to natural sand in normal concrete. The study mainly focusses on understanding the influence of different replacement ratios and suitability of slag replacement ratios higher than 50% in normal concrete. The blast furnace slag used in the present study was processed using a vertical shaft impactor to modify the particle shape and enhance its aggregate properties. The replacement was done at 50%, 75% and 100% by mass of natural sand in the concrete mix. The concrete properties such as slump, compressive strength, splitting tensile strength and flexural strength, were studied at all the replacement ratios.

1. Introduction

The acute shortage of natural sand has made the Indian construction industry to look for all possible alternative materials. In the recent past there are many studies reported on using materials such as waste foundry sand, coal bottom ash, copper slag, steel slag, fly ash, recycled concrete aggregates and mined mineral tailings [1–3], as a replacement to natural sand in concrete. Among all the possible alternative materials explored, metallurgical slags have drawn the attention of researchers and are being examined for their suitability as fine aggregate in concrete. The lack of appropriate engineering data about such materials makes it hard for the designers to use them optimally in construction. In India, a major share of granulated blast furnace slag (GBS) is consumed by the cement industries. Whenever the cement production slows down, the continuously operating blast furnaces make a huge pile of slag and the stocks of this excess slag are mostly dumped on the ground covering fertile land spaces. Researchers are exploring ways to utilise GBS as a replacement to fine aggregates in mortars and concretes. Several investigations are conducted so far to characterize GBS and its influence on properties of concrete [1,4,5]. Yuksel and Genc [6], studied non ground blast furnace slag to make concrete and have reported that compressive strength and flexural strength decreased at higher replacement ratios. Most of the investigations [1,4–6] recommend a GBS replacement ratio in the range of 30%-60%, stating that at higher replacement levels the workability, compressive and tensile strengths of concrete were negatively affected. Nevertheless, attempt to process GBS to improve its material properties and to study the suitability of processed GBS as fine aggregate in concrete have been less explored.

2. GBS and the processed granulated blast furnace slag

In the present study, at first, unprocessed GBS sample sourced from a local steel plant was investigated (Jindal Steel works, Vijayanagar, India). During the preliminary investigations it was observed that unprocessed GBS had less bulk density, sharp edges and rough surface texture, giving rise to serious workability issues. The material had large number of sharp glassy particles making it extremely difficult to handle during the making of concrete. Further, the GBS was processed at steel plant using a vertical shaft impactor to modify the shape and texture of GBS. The processing method adopted improved appearance, shape and texture of particles. The breakage of particles during processing resulted in increase in the percentage of finer fraction improving the bulk density of aggregate. The typical shape and morphology of GBS and processed GBS (PGBS) observed under scanning electron microscope is shown in figure 1. The properties of GBS, PGBS and natural sand are provided in table 1.

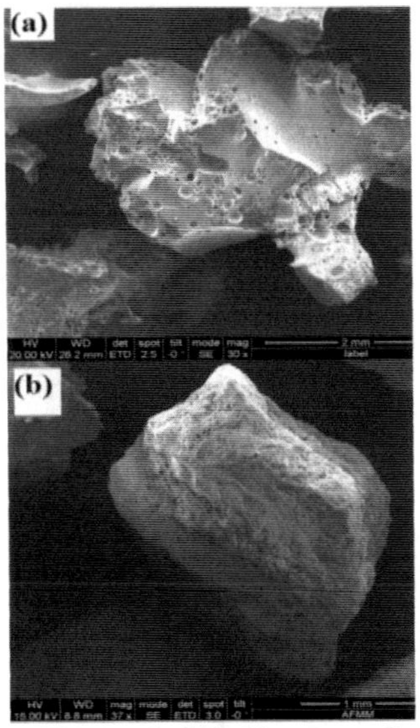

Figure 1. SEM images of (a) GBS and (b) PGBS particle.

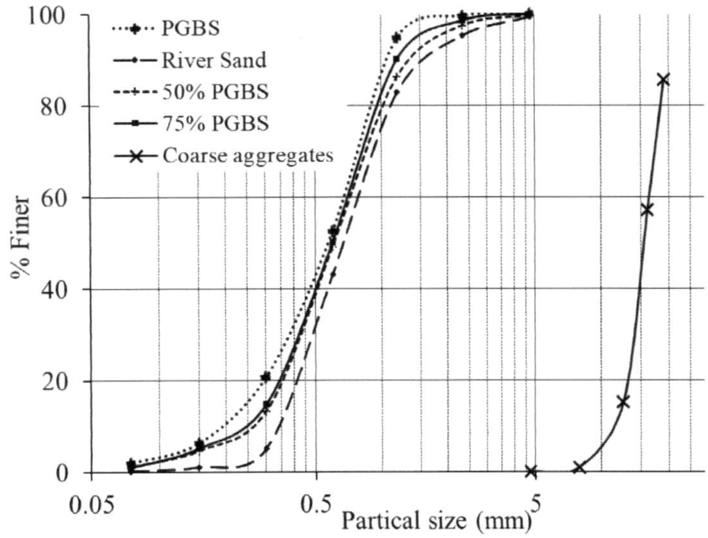

Figure 2. Particle size distribution of aggregates.

Table 1. Properties of GBS, PGBS and natural sand.

Material properties	GBS	PGBS	Natural sand
Bulk density, kg/m^3	1190	1340	1450
Specific gravity	2.65	2.71	2.66
Water absorption, (%) by mass	4.81	5.84	1.07
Fineness modulus	2.89	2.40	2.71

3. Scope of work and experimental program

The present study aimed at characterising the physical and chemical properties of PGBS aggregates. Further, concrete mixes with three levels of PGBS replacement were examined. The mix having 100 % natural sand as fine aggregate was considered as control concrete. Other concrete mixes were prepared by replacing natural sand at 50 %, 75 % and 100 % by PGBS in the control mix. The details of concrete mix and the mix designations are provided in table 2. The investigation aimed at studying the influence of natural sand to PGBS replacement ratios on the properties of concrete such as workability, compressive strength, splitting tensile strength and modulus of rupture. The study focussed on examining compressive strength of concrete mixes at 7, 28 and 90 days and tensile strengths of concrete mixes at 7 and 28 days. The results obtained for concrete mixes with different PGBS content were compared with control concrete mixes of similar age.

4. Materials and mix proportions

Ordinary portland cement (OPC) conforming to Indian Standard specification IS:12269 [7] with a consistency of 32 % and a specific gravity of 3.15 was used. The cement showed an initial setting time and final setting of 165 and 291 minutes respectively and 28-day compressive strength of 56.2 MPa. The physical properties of natural sand and PGBS are presented in table 1. Along with PGBS aggregate the natural sand conforming to IS:383 [8] was used in this study. The particle size distribution of fine aggregates used in the study is shown in figure 2. The PGBS had 33.6% SiO_2, and 22.3% Al_2O_3 along with 26.3% CaO. The MgO in PGBS was about 5% with minor quantities of other oxides. The PGBS showed a lime reactivity of 0.24 MPa. The crushed stone aggregate of size 20 mm and down was used as coarse aggregate. It had a specific gravity of 2.71, bulk density of 1580 kg/m3, and a water absorption 0.35%. Potable water was used to prepare the concrete mixes. A Chloride free, sulphonated naphthalene polymer based super plasticizer from Fosroc brand complying with ASTM C494 standard [9] was used. The concrete mix was designed following the guidelines of IS:456 [10] and IS:10262 [11]. The design was aimed at yielding a concrete slump in the range of 50 to 100 mm. It also aimed at a concrete characteristic compressive strength (f_{ck}) of 20 MPa and a corresponding target cube compressive strength of at least 26.5 MPa at 28 days as per IS:516 [12]. Several trial mixes were studied to understand the relationship of w/c to slump before finalizing the concrete mix proportions. The super plasticizer dosage was adjusted such that the concrete with 100% PGBS yielded a minimum slump of 50 mm. The fine aggregate and coarse aggregate were in air-dried condition at the time of mixing. The moisture content of each batch of aggregates were measured before weigh batching and the total water was adjusted accordingly before adding water to the concrete mix. The details of the concrete mix proportions are summarised in table 2.

5. Specimen preparation and testing

The concrete was mixed, placed, compacted and finished as per the guidelines of IS:516 [12]. A motorized pan mixer was used to mix the concrete ingredients. The consistency of fresh concrete was determined immediately after mixing, by measuring slump as per IS:1199 [13]. All the specimen moulds were compacted using a needle vibrator as specified in IS:516 [12]. The compressive strength was determined by crushing concrete cubes of 150 mm. The concrete cylinders having 150 mm diameter and 300 mm height were prepared for split tensile strength determination. Unreinforced concrete prisms of dimension 100 mm x100 mm x500 mm were cast for flexural strength determination. All the test specimens were cured by soaking in clean water until the date of testing or up to 28 days from the time of casting.

Table 2. Details of concrete mixes.

Mix designations		A	B	C	D
Cement	(kg/m³)	320	320	320	320
Natural sand	(kg/m³)	870.4	435.2	217.6	0
PGBS	(kg/m³)	0	435.2	652.8	870.4
PGBS replacement ratio (% by mass)	(%)	0	50	75	100
Coarse Aggregate	(kg/m³)	1024	1024	1024	1024
W/C ratio	-	0.5	0.5	0.5	0.5
Water content	(kg/m³)	160	160	160	160
Super Plasticizer	(kg/m³)	3.8	3.8	3.8	3.8
Slump	(mm)	112	82	73	58

6. Results and discussions

The results of tests conducted on concrete mixes such as slump measurement, compressive strength, split tensile strength, flexural strength studies are discussed in the following sections.

6.1. Concrete workability

The properties of fine aggregate in concrete have a direct influence on the workability of concrete. The shape and surface texture of the fine aggregate affects the properties of fresh concrete more than the hardened concrete. Compared to smooth and rounded aggregates, rough, angular and elongated particles need higher cement paste to produce desired workability [14]. The slump of fresh concrete was measured within 15 minutes of mixing. Table 2 provides the slump measurements obtained for each concrete mix. The mix with 100% natural sand had the highest slump, i.e., 112 mm. The slump reduced with increasing replacement levels of PGBS. The mix with 100% PGBS gave the least slump compared to all other mixes. The slump loss with 100% PGBS replacement was nearly 50 percent. The PGBS aggregate were more angular, rough and showed higher water absorption compared to natural sand. Therefore, with increase in the PGBS content, the effort required to overcome the intergranular friction also increased, resulting in reduced workability.

6.2. Compressive strength (f_c)

The compressive strength of concrete mixes with respect to PGBS content, at 7 days, 28 days and 90 days has been shown in figure 3. All specimens were cured as mentioned in section 5. The concrete specimens tested at 90 days were cured in water for 28 days and were stored at room temperature until the date of testing. At 7 days, mix B and C showed 7% and 5% higher compressive strength, respectively, than control mix, while mix D with 100% slag showed a reduction in strength of 3%. All concrete mixes achieved the target mean compressive strength at 28 days as per the mix design. At 28 days, the compressive strengths of PGBS based concrete mixes was higher compared to that of control mx. The percentage variation of compressive strength of PGBS based concrete mixes with respect to the control mix is shown in figure 4. Mix 'B' with 50% PGBS replacement ratio performed better under compression compared to mix C and D at all testing ages. The higher compressive strength of mix B could be due to two reasons. One, the inclusion of finer particles of PGBS resulted in better particle distribution and the pozzolanic activity of PGBS aggregate. The PGBS had 17% higher finer fraction compared to natural sand. The combination of 50% of PGBS and natural sand by mass improved the particle distribution in finer zone of aggregate (figure 2). The lime reactivity of PGBS as mentioned in section 3 indicates that, though very weak, PGBS particles may still show some pozzolanic reactions with time. Therefore, there could be some amount of pozzolanic activity shown by finer slag particles contributing to the formation of additional C-S-H. However, quantification of hydration products of concrete mixes and more thorough study of concrete microstructure may help to decisively comment on the compressive strength variation in concrete mixes.

6.3. Splitting tensile strength (fs) and flexural strength (ft) of concrete mixes

The variation of splitting tensile strength and flexural strength of concrete mixes with respect to the PGBS replacement ratios is shown in figure 5 and figure 6 respectively. The variation in the split tensile strength and flexural strength with respect to PGBS content closely followed the trend observed in the case of compressive strength. At 7 days mix 'D' with 100% PGBS achieved lower splitting tensile strength and flexural strength than control mix. At 28 days, the flexural strength of all PGBS based mixes were higher than that of control mix. Mix 'B' with 50% PGBS showed highest split tensile strength and flexural strength compared to all other mixes. According to IS:456 [10], the flexural strength of normal concrete is considered as $0.7\sqrt{f_{ck}}$, where f_{ck} is the characteristic compressive strength of concrete and with that criteria, all the PGBS concrete mixes in the present study achieved the minimum expected flexural strength. For normal concrete, the ratios of splitting tensile strength to compressive strength (fs/fc) and the ratio of flexural strength to compressive strength (ft/fc) ranges between 7-9% and 11-18% respectively [14]. Also, the flexural strength of concrete is almost 1.5 to 2 times that of splitting tensile strength. provides the (fs/fc), (ft/fc) and (ft/fs) values obtained for PGBS concrete mixes. It appears that these values obtained for concrete mixes considered in the present study (table 3) are within the range of values expected for normal concrete. The surface texture and gradation of fine aggregate plays an important role in influencing the crack development of concrete under tension. The rough textured aggregates improves the tensile behaviour of concrete compared to more rounded aggregates during early age of concrete [15]. In this present case, the rough texture of PGBS aggregate particles may have influenced the formation of stronger physical bond between cement paste and aggregate, therefore improving the tensile behaviour of concrete.

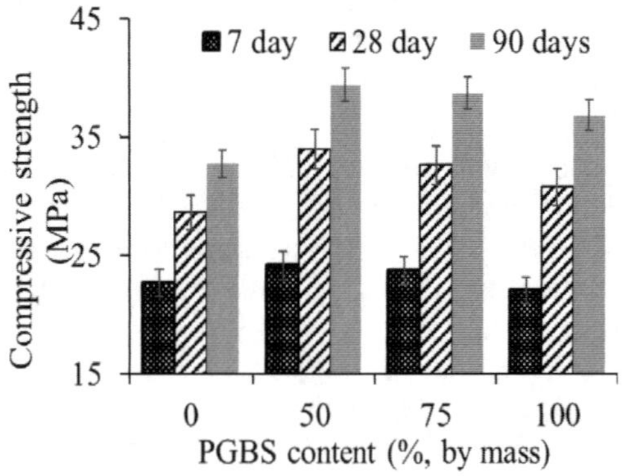

Figure 3. Compressive strength of concrete mixes.

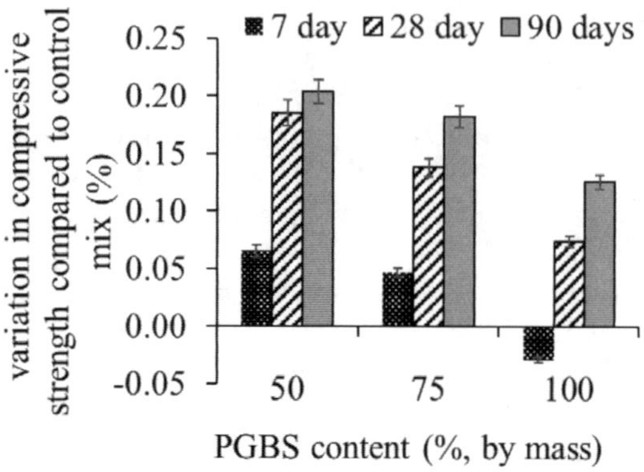

Figure 4. Variation of compressive strengths with respect to control mix.

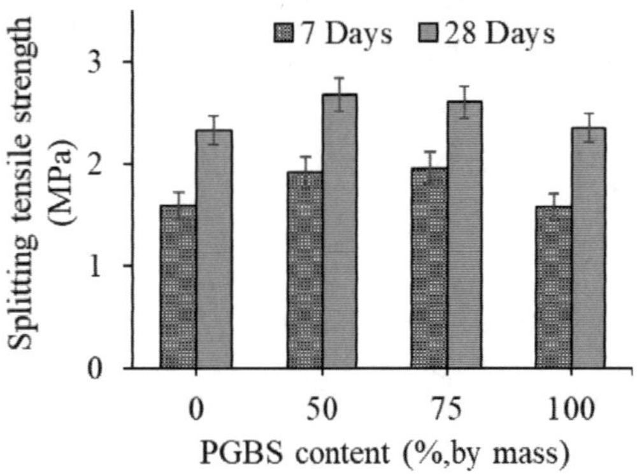

Figure 5. Splitting tensile strength of concrete mixes.

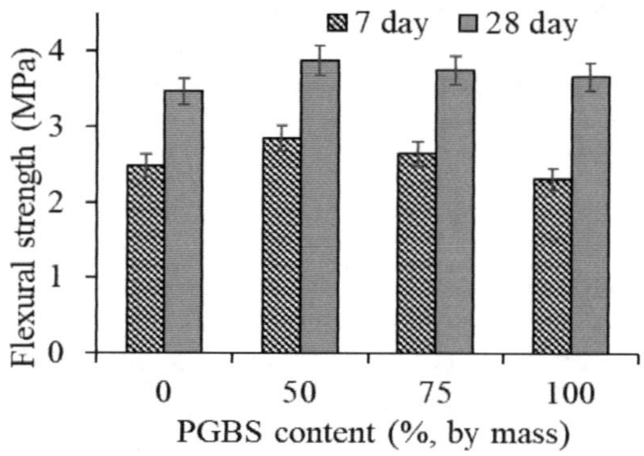

Figure 6. Variation of flexural strength of concrete mixes.

Table 3. The ratios of (f_s / f_c), (f_t / f_c) and (f_t / f_s), of concrete mixes.

PGBS content (%, by mass)	(f_s / f_c) %		(f_t / f_c) %		(f_t / f_s)	
	7 days	28 days	7 days	28 days	7 days	28 days
	7.0	8.0	11	11	1.57	1.50
50	8.0	7.8	12	11	1.49	1.45
75	8.2	8.0	11	12	1.35	1.44
100	7.1	8.1	11	12	1.47	1.58

7. Conclusions

In the present study it was observed that addition of PGBS in place of natural sand positively influenced the properties of concrete except workability and shrinkage. Although, the concrete with 50% replacement levels gave the highest numbers considering strength under compression and tension, it can be concluded that even at 100 % replacement levels, the PGBS concrete showed satisfactory mechanical properties. Therefore, both partial and complete replacement of natural aggregates in concrete is possible with PGBS. Replacement of PGBS in the concrete reduced the workability of concrete significantly. The workability of the concrete decreased with increase in PGBS replacement percentages. However, with incorporation of suitable modern-day superplasticizers, one can achieve desired workability and make the PGBS concrete suitable for the desired application. The PGBS concrete met the strength requirements as per IS 456, both at 7 days and 28 days, at all replacement levels. However, 50% and 75% PGBS replacement showed higher compressive strength, splitting tensile strength and flexural strength when compared to 100% replacement with PGBS.

8. References

[1] Yüksel I, Özkan Ö and Bilir T 2006 Use of granulated blast-furnace slag in concrete as fine aggregate *ACI Mater. J.* **103** 203–8
[2] Pellegrino C and Gaddo V 2009 Mechanical and durability characteristics of concrete containing EAF slag as aggregate *Cem. Concr. Compos.* **31** 663–71
[3] Brand A S and Roesler J R 2015 Steel furnace slag aggregate expansion and hardened concrete properties *Cem. Concr. Compos.* **60** 1–9
[4] Yüksel I, Bilir T and Özkan Ö 2007 Durability of concrete incorporating non-ground blast furnace slag and bottom ash as fine aggregate *Build. Environ.* **42** 2651–9
[5] Lee K M, Lee H K, Lee S H and Kim G Y 2006 Autogenous shrinkage of concrete containing granulated blast-furnace slag *Cem. Concr. Res.* **36** 1279–85
[6] Yüksel I and Genç A 2007 Properties of concrete containing nonground ash and slag as fine aggregate *ACI Mater. J.* **104** 397–403
[7] IS:12269 2013 Specification for 53 grade ordinary portaland cement, *Bur. Indian Stand. New Delhi*
[8] IS:383 2002 Specification for Coarse and Fine Aggregate From Natural Sources for Concrete *Bur. Indian Stand. New Delhi*
[9] ASTM C 494 2019 Standard specification for chemical admixtures for concrete *ASTM Int.*
[10] IS:456 2000 Plain and Reinforced Concrete code of practice *Bur. Indian Stand. New Delhi*
[11] IS:10262 2009 Concrete mix proportioning guidelines *Bur. Indian Stand. New Delhi*
[12] IS:516 2004 Methods of tests for strength of concrete, *Bur. Indian Stand. New Delhi*
[13] IS:1199 1999 Methods of sampling and analysis of concrete *Bur. Indian Stand. New Delhi*
[14] Mehta P K and Monteiro P J M 2014 *Concrete - microstructure, properties, and materials* (New Delhi: Mc Graw Hill)
[15] Neville A M 2011 *Properties of Concrete* (New Delhi: Dorling Kindersley (India) Pvt. ltd)

Acknowledgements

Authors wish to sincerely thank Ministry of Human Resources and Development, Government of India, and Indian Institute of Science Bangalore, for financially supporting and facilitating this research work.

IOP Conf. Series: Materials Science and Engineering **829** (2020) 012009 doi:10.1088/1757-899X/829/1/012009

Exploring the Mechanism of Residential Attractiveness in Compact Urban Areas – a Case Study of Hong Kong

Yunxi Bai[1,5], Jusheng Song[1,6], Shanshan Wu[2], S M Lo[3] and Gang Bai[4]

[1]School of Architecture, Harbin Institute of Technology, HIT Campus of University Town of Shenzhen, Shenzhen, Guangdong, China
[2] Department of Architecture, Tongji University, Siping Road, Shanghai, China
[3] Department of Architecture and Civil Engineering, City University of Hong Kong, Tat Chee Avenue, Kowloon, Hong Kong
[4] College of Computer and Control Engineering, Nankai University, Weijin Road, Tianjin, China

[5]yxbai2-c@my.cityu.edu.hk, [6]songyuangc@163.com

Abstract. Compact cities adopt high density, mixed land use and transit-oriented development (TOD). For sustainable development, compact development reduces the use of automobile and lessens the overdevelopment of land resources. For solving urban sprawl and housing shortage, more and more metropolises have adopted compact development. Academics, planners and policy makers mainly focused on evaluating the impacts of neighborhood attributes on housing prices and residential movements. Few have examined the relationships between residents' demand and residential attractiveness in compact cities. Therefore, this study develops Residential Attractiveness Evaluation (RAE) Model to evaluate the attractive degree of urban areas based on residents' demand in compact cities. Hong Kong is chosen as the study area which is a typical compact city. The results show that districts near the CBD and both sides of Victoria Harbor have highest residential attractiveness than other areas. Community planners and policy makers should consider residential attractiveness based on residents' demand in efforts to promote sustainable development. Residential attractiveness is the foundation of residential land use planning and policies. Planners in compact cities should adopt various planning strategies according to different residential attractiveness. Moreover, residential attractiveness is the benchmark of evaluating residents' satisfaction of new residential planning and policies.

1. Introduction

For rapid urbanization, more and more cities have faced many urban problems, like traffic congestion, environment pollution and urban sprawl [1]. The low-density urban development based on the automobile which has been widely adopted in European and North American countries is not appropriate in Asian cities. Thus compact development was presented and popular with urban planners and policy-makers [2]. The characteristics of compact development are high density, mixed land-use and transit-oriented development [3, 4]. Compact cities efficiently use land resource and decrease the waste of land. They can protect the habitat environment and wild landscapes. In addition, compact urbanization can promote public transit development and reduce the use of cars. It aims at protecting environment and developing sustainable communities. For the success of compact development, urban planners and policy makers should provide good living environment and increase the attractiveness of residential areas for residents. They should understand the attractiveness of residential areas based on residents' demand in compact cities, and provide sustainable development strategies according to various conditions.

Architects, planners and economists focused on the influences of housing attributes on housing prices and residential mobility. Previous studies indicated that some attributes influenced residential movement and behavior, like household and neighborhood characteristics, accessibility factors, and economic attributes. Residents moved to a new place which can provide best combination of influential attributes [5]. Although extensive literature has indicated the relationships between housing attributes and residential mobility [6, 7], the spatial attractiveness of each area according to residents' demand has been rarely evaluated. In addition, Extensive literature paid more attention on residential mobility in low density and automobile commuting cities [8, 9]. They have limitations in capturing the residents' demand on high density and public transit-oriented cities, like Hong Kong. Therefore, this work attempts to develop a systematic framework to evaluate the spatial attractive degree of compact urban areas based on residents' demand. This work can help urban planners and policy makers to understand residential attractiveness of each district in planning process. They should provide various planning strategies according to different attractive degrees to promote sustainable development.

Section 2 introduces the characteristics of study areas in Hong Kong. Furthermore, section 3 describes the process of data collection and the development of systematic framework to measure attractiveness of residential urban areas based on residents' demand. Then section 4 shows how to apply the systematic framework for evaluating the residential attractiveness of urban areas in Hong Kong. Section 5 discusses the application for supporting sustainable land use planning and policies. Finally, section 6 presents the contributions, and points out the limitations and future research.

2. Study area

Hong Kong is one of famous compact cities which is located in the south of China. In recent decades, urban planners and policy makers have adopted TOD and compact development in Hong Kong which can promote it more sustainable [2, 4]. Hong Kong has strict planning and policy restrictions on the development of new urban land. These policies can protect wild landscapes and our habitat, and promote sustainable development. Hong Kong government has used the mass transit railway (MTR) as the backbone of the public transport system. It can reduce the use of automobile and air pollution. Hong Kong has also developed densely populated buildings and mixed land use to reduce the waste of resources. Hong Kong is composed of three major areas: Hong Kong Island, Kowloon and New Territories. In addition, the CBD of Hong Kong is located in Hong Kong Island. From the perspective of town planning, the whole territory of Hong Kong is categorized by the Planning Department into three levels: Primary Planning Units (PPUs), Secondary Planning Units (SPUs) and Tertiary Planning Units (TPUs). TPU is the smallest district unit which has the available data from Population Census in Hong Kong. It is appropriate to be regarded as the unit of residential district in this work. Therefore, this study evaluates the attractiveness of TPUs for residents in Hong Kong.

3. Materials and methods

3.1. Analytical framework

This research develops a systematic framework to evaluate residential attractiveness of urban areas for residents in content of compact urbanization. Based on the review of theories about residential mobility, Residential Attractiveness Evaluation (RAE) Model which is derived by Multiplicative competitive interaction (MCI) model is developed to estimate residential attractiveness of urban areas according to residents' demand. Three attributes which significant influence residents' preferences in compact cities are identified as independent variables, including access to metro station, access to CBD and median household income. Two types of revealed-preference data consisting of geographical data and socio-demographical data are collected. In this research, residential attractiveness of each urban area in Hong Kong is measured to illustrate how to analyze the relationship between residents' demand and residential attractiveness in compact cities.

3.2. Variables

Residents consider some location attributes when they decide to live in a residential area, like accessibility, neighborhood and socio-economic characteristics [7, 10]. Compact urbanization is famous by public TOD, high-density and mixed land use [3]. In compact cities, there are three attributes which have significant impacts on residential mobility, access to the metro station (METRO), access to the CBD (CBD) and median household income (MHI) [11, 12].

In compact cities, the public chose metro as the initial travel mode, including going to work, study or shopping [11]. Thus, access to the nearest metro station plays a significant role in the residential movement and behavior in compact cities [12]. Accessibility of the CBD has also been proved as an important attribute for residents in compact cities [7, 12]. It has been aggregated with sufficient and abundant public facilities, and has a great attraction for the public. Moreover, median household income decides the affordable value of residential properties. It has been regarded as an important attribute which influenced the residents' willingness to move [5, 6, 12]. Therefore, access to metro station, access to CBD and median household income are regarded as significant attributes which are used in the RAE model.

3.3. Data collection

This research develops an evaluation method to measure the attractiveness of all available areas for residents in a compact city. The data of 289 TPUs in Hong Kong are collected to test the proposed model. Figure 1 presents the location of TPUs in Hong Kong. There are three variables in the proposed model: network distance to the nearest metro station, network distance to the CBD, and median monthly household income. The location of all TPUs are from Hong Kong Population Census Database. Meanwhile, the coordinate of CBD and metro stations are processed from Open Street Map to ArcGIS. The data of distance are easier to collect and more accurate than the data of traveling time in this research. So network distance from central point of TPUs to the nearest metro station and to the CBD are measured by Google Map. Moreover, the data of median household income are from HK Population Census Database. However, TPU 251, 434, 814, 834, 913, 933, 934, and 963 which do not have available data are removed. The descriptive statistics of variables are presented in table 1.

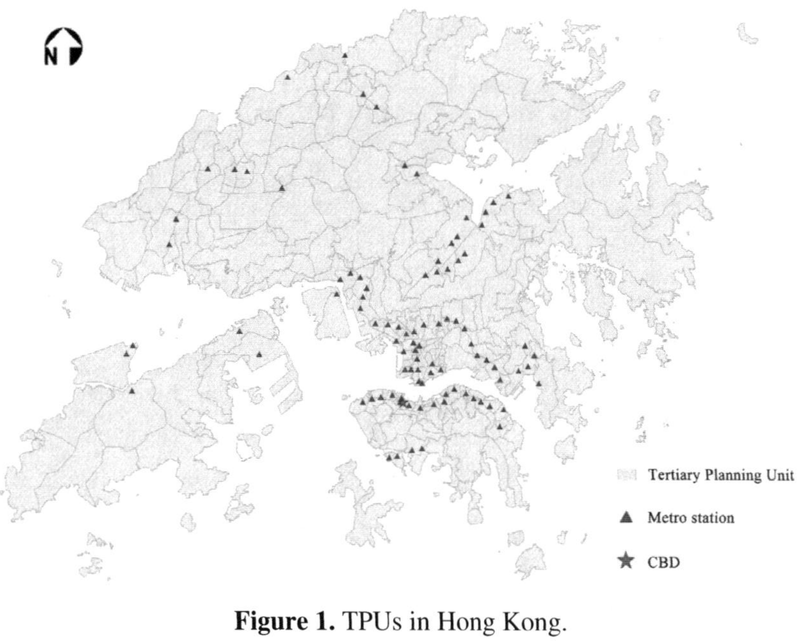

Figure 1. TPUs in Hong Kong.

Table 1. Descriptive statistics of major variables.

Variable	Minimum	Maximum	Mean	Std. Deviation
METRO	0.07	27.90	4.3855	5.26065
CBD	0.10	59.50	19.8700	13.32573
MHI	4778.00	208293.00	42122.1750	38149.37021

3.4. Residential attractiveness evaluation model

In this research, the Residential Attractiveness Evaluation Model is proposed to evaluate the attractiveness of areas for residents in Hong Kong. Discrete choice theory was always used to predict the optimal choice between two or more discrete alternatives for a rational consumer based on his/her preference [10]. MCI model is derived by discrete choice theories and used to measure the attractiveness of a product for consumers [13, 14]. Residential properties can be regarded as a product which are purchased by residents. Therefore, this study adopts MCI model to evaluate residential attractiveness (RA) of districts in compact cities. The RAE model is derived by MCI model which is presented as follow:

$$RA_i = exp(\alpha_i) \cdot \prod_{k=1}^{K} X_{ki}^{\beta_k} \cdot \varepsilon_i \quad (1)$$

where i is the number of a district in a compact city. RA_i is the residential attractiveness of the ith district. α_i is the intrinsic component. K is the number of set of attributes, while k is the number of an influenced attribute ($k \in K$). Moreover, X_k is the kth attribute that significantly influences residential mobility. β_k is a parameter of kth attribute, and ε_i is an error term.

In Hong Kong, three attributes have most significant influences on residents' preferences when they want to move in a residential area. Independent variables are *Access to Metro Station (METRO)*, *Access to CBD (CBD)* and *Median Household Income (MHI)*. The proposed model is presented as follows:

$$RA_i = exp(\alpha_i) \cdot (METRO_i^{\gamma} \cdot CBD_i^{\lambda} \cdot MHI_i^{\mu}) \cdot \varepsilon_i \quad (2)$$

γ, λ, and μ are the parameter of major attributes in the proposed model. In Hong Kong, an exponential parameter γ for network distance to the nearest metro station is 1. A distance decay parameter λ for network distance to CBD is 3, and an exponential parameter μ for median household income is 1. RA can be used for urban planners and policy-makers to understand the attractive degree of each district for residents in compact cities, and to support sustainable residential land use planning and policies.

4. Results

Residents have different attitudes to residential properties in different districts according to their location and neighborhood characteristics. So, each area has different attractiveness for residents in a compact city. This study develops the RAE model to evaluate attractive degree of each district for the public in Hong Kong. Quantiles are used to divide the dataset into contiguous intervals according to the range of a probability distribution. This work selects decile categorization which is appropriate to present the residential attractiveness of each TPU in Hong Kong.

Figure 2 shows a visual map to directly present residential attractiveness of each TPU in Hong Kong. Results are divided into 10 parts. Each level has different attractiveness of these districts. Level 1 represents that a few residents are willing to move into this district, while Level 10 means the opposite. More residents want to move in the district in a higher level. The residential attractiveness degree map of TPUs in Hong Kong indicates that citizens in Hong Kong prefer to live in areas near the CBD and both sides of Victoria Harbor. In general, residential properties in Hong Kong Island attract more consumers to purchase than ones in Kowloon which are more attractive than those in New Territories.

IOP Conf. Series: Materials Science and Engineering **829** (2020) 012009 doi:10.1088/1757-899X/829/1/012009

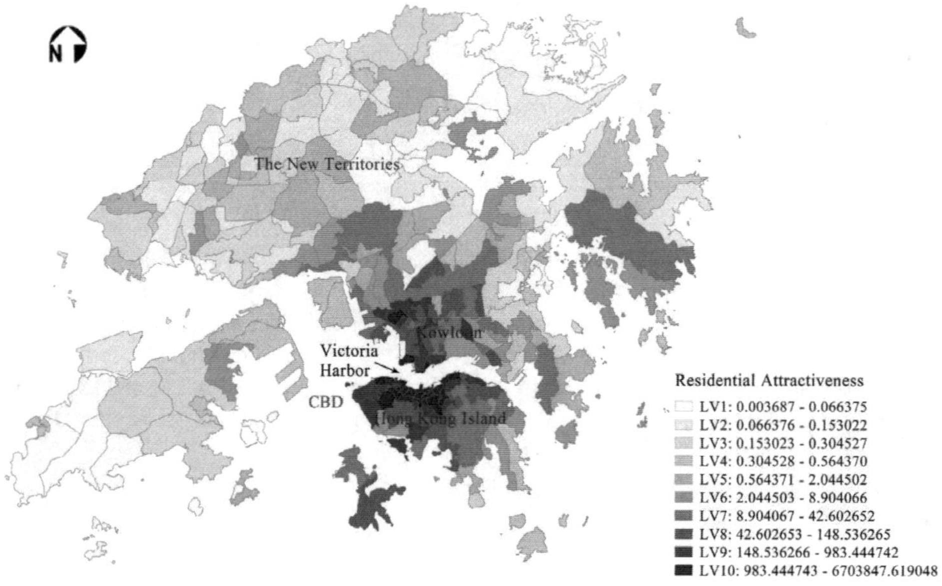

Figure 2. Residential attractiveness degree map of TPUs in Hong Kong.

5. Discussion

Compact cities are famous by TOD, mixed land use and high-density development. Compact development not only provides convenient and good services for citizens' residence and travel, but also protects our habitat and supports sustainable development. For sustainable development, residential land use planning and policies should meet current demand of residents in compact cities [15, 16]. Therefore, the evaluation of residential attractiveness can help urban planners and government workers to understand living level of each district in their cities. They should adopt different planning strategies according to different attractive degree of districts for sustainable development.

By using the RAE model, this study evaluates the residential attractiveness of urban areas based on residents' demand in compact cities. Residential attractiveness can be used in land use planning and policies for supporting sustainable development. First, residential attractiveness can be a benchmark of residential level of districts based on residents' preferences in compact cities. Urban planners should adopt different planning strategies for urban areas with different attractive degree. The districts with high attractiveness can generate a large number of people. Planning department should consider walkable and transit-oriented development in these areas to develop sustainable neighborhoods. Furthermore, planners and policy makers should provide sufficient infrastructure and facilities for districts with lower residential attractiveness. They should also promote public transit-oriented development in these districts. Rapid transit investments can attract citizens to move in these districts and reduce the pressure on the population of city center. It also decreases travel costs and pollution emissions to improve sustainable development [3].

Second, Residential attractiveness can also help urban planners to decide whether the dominant land use is residential in land use planning. It can be used by policy makers to evaluate residents' satisfaction with new land use policies and residential developments. Government officers can measure the residential attractiveness of a specified area before and after the implementation of new policy and planning. The change of RAs reflects whether the new policies meet the needs of residents for living environment. Third, the RAE model would be a better method to gauge endogenous linkage between residents' demand and residential attractiveness in other compact cities, like Tokyo, Soul, and Shanghai. It can also be developed to further evaluate citizens' satisfaction for other types of land use, like shopping centers, working places, public facilities, etc. New land use planning and policies with the help of these methods can provide attractive and comfortable neighborhood for residents to support sustainable development in compact cities.

6. Conclusions

Recent literature has mainly focused on investigating the influences of housing attributes on residential location choice and housing prices. However, whether the urban areas meet the demand of residents in compact cities has rarely been evaluated. Therefore, this study develops the systematic framework to evaluate the attractiveness of urban areas for residents in compact cities for supporting sustainable residential land use planning and policies. The RAE model which is derived by the MCI model is developed to evaluate attractive degree of areas for residents in compact cities. Furthermore, this research shows how to apply the RAE model for estimating residential attractiveness of available TPUs based on residents' demand in Hong Kong. The findings indicate that residential properties in CBD and both sides of Victoria Harbor can attract more residents to live in. Citizens prefer to live in Hong Kong Island to Kowloon and New Territories. Urban planners and policy makers should take residential attractiveness into account when they undertake public infrastructure and transit investments. It is also the foundation of sustainable residential land use planning and policies. This study can help the government to evaluate attractive level of areas for residents in compact cities. Planners should present appropriate planning strategies for various areas with different residential attractiveness. It can also be adopted to estimate the residents' satisfaction of new land use planning and policies.

There are some limitations of this study. This study collects the data of Hong Kong to illustrate how to evaluate the residential attractiveness of urban areas based on residents' demand. However, major variables and parameters may be different in other compact cities. More empirical data from other regions should be collected to test in further work. Moreover, this research collects the geographic data by measuring the network distance from the central point of each TPU to other locations. But, large TPUs should be divided into some parts which have different location attributes and different attractiveness for residents. The residential attractiveness of more location in TPUs will be measured in future research and the results will be more accurate.

7. References

[1] Obregón-Biosca S A, Romero-Navarrete J A, Mendoza-Sanchez J F, Betanzo-Quezada E 2015 *J. Urb. Plan. Deve.* **142** 05015005
[2] Cervero R, Murakami J 2009 *J. Urban Studies* **46** 2019–43
[3] Houston D, Boarnet M G, Ferguson G, Spears S 2015 *J. Urban Studies* **52** 938–59
[4] Liao F H, Farber S, Ewing R 2015 J. Urban Studies 52 314–37
[5] Earnhart D 2001 *J. Land econom.* **77** 12–29
[6] Paleti R, Bhat C, Pendyala R 2013 *J. Trans. Res. Rec.* **2382** 162–72
[7] Kim J H, Pagliara F, Preston J 2005 *J. Urban Studies* **42** 1621–36
[8] Cervero R, Wu K L 1997 *J. Envir. Plan.* A **29** 865–86
[9] Mok D 2007 *J. Urban Studies* **44** 723–50
[10] Greene M, de Dios Ortúzar J 2002 *J. Int. Plan. Stud.* **7** 55–87
[11] Tse R Y, Love P E 2000 *J. Prop. Manag.* **18** 366–74
[12] Inoa I A, Picard N, De Palma 2015 *J. Math. Popu. Stud.* **22** 4–36
[13] Gonzalez-Benito O, Munoz-Gallego P A, Kopalle P K 2005 *J. Retail.* **81** 59–73
[14] Wu S S, Kuang H, Lo S M 2018 *J. Urb. Plan. Deve.* **145** 04018047
[15] Hansson A G 2017 *J. Land use poli.* **64** 470–78
[16] Yu K, Hui E 2018 *J. Land use poli.*, **72** 341–53

Conceptual Residential Design Framework to enhance Well-being of Elderly in Thailand

Porntip Ruengtam

Faculty of Architecture Urban and Creative Arts, Mahasarakham University, Mahasarakham 44150 Thailand

porntip.r@msu.ac.th

Abstract. A conceptual residential design framework was modelled using healing environmental and universal design concepts to enhance the well-being of the elderly in Thailand. Research methods involved informal interviews and observations for three Thai case studies. Interview results were summarized with related theories and literature. All factors were collated and listed. Seven experts in related fields were employed to confirm the factors and framework following the Delphi technique. Results showed that elderly characteristics had a positive direct effect on requirements of the elderly in a residential community which comprised two sub-factors including (1) healing environment, and (2) universal design. Each factor affected the willingness of the elderly to join communal events of the residential community. This framework will be used for quantitative analysis to determine future policy for environmental design and management of residential communities to enhance the well-being of the elderly in Thailand.

1. Introduction

An annual report of the Thai elderly [1], defined as those who are 60 years of age and above, concluded that they will account for 19.1%, 26.6%, and 32.1% of the Thai population in the years 2020, 2030, and 2040 respectively. This prediction has begged the question how the Thai government will cope when these situations arise. In response, a resolution has been proposed to enhance the well-being of the elderly residential community through healing and universal design concepts, but what should the environment look like? The objective of this research is to study and formulate a conceptual research framework of characteristics of the Thai elderly and their requirements for healing, and the universal design of the residential community. In order to achieve this objective, this research has: Identified components of the requirements of the Thai elderly, and their measurements. Developed a conceptual framework that demonstrates the relationships between the characteristics of the elderly, their requirements, and their willingness to join the residential community. The framework of characteristics will be proved by collecting empirical data used for quantitative analysis. The outcome will be a policy for the design and management of the residential community for enhancing the well-being of the elderly in Thailand.

2. Literature review

The National Committee for the Elderly, under Thailand's Ministry of Social Development and Human Security [2], has defined the elderly as able to utilize social development resources, entitling them to recognition and support of their family, community, and state to lead a valuable life with dignity, and sustain their health and living standards as long as possible.

WHO (World Health Organization) [3] has stated on their website that most developed countries in the world have accepted the chronological age of 65 years as a definition of elderly or older people, but like many westernized concepts, this does not adapt well to the situation in Africa. While this definition is somewhat arbitrary, it is usually associated with the age at which one can begin to receive pension benefits. At the moment, there is no UN (United Nations) standard numerical criterion, but the UN-agreed cut- off age is 60+ to refer to the older population. In this research, the Thai elderly is also defined by the age of 60+, together with their health condition, classified in 5 levels, as follows: (1) Strong: can help themselves with everything, (2) Sometimes need some assistance, (3) Usually need assistance, (4) Need constant assistance, and (5) Unable to move; need constant assistance.

Plunz [4] stated that environmental design is the process of addressing surrounding environmental parameters when devising plans, programs, policies, buildings, or products. Environmental design can also refer to the applied arts and sciences - dealing with creating a human-designed environment. These fields include architecture, geography, urban planning, landscape architecture, and interior design. Environmental design studies the physical surroundings that provide the setting for human activity, ranging in scale from buildings and parks, green spaces for neighborhoods, and the local community. Environmental design is defined as the physical and constructed environment in which people live, work, and enjoy recreation on a day-to-day basis. In addition, environmental design is concerned with the way these places are experienced and used, as well as other aesthetic elements that contribute to the quality of community environments. Environmental design would be considered in this research in terms of requirements for the Thai elderly.

WHO [5] has given an initial definition of "health" in a common description. In 1998, the definition was improved to mean a dynamic state of four-dimensions, including physical, mental, social, and spiritual well-being. Jonas and Chez [6] believed an emphasis on healing is key to the future medical management of chronic illness and the establishment of sustainable approaches to health care. Defined as the process of recovery, repair, and return to wholeness, healing is the foundation for a vision of medicine that integrates diverse approaches from around the world for the alleviation of suffering, the enhancement of well-being, and the treatment of chronic illness.

Healing is facilitated through the development of proper attitudes and intentions in both the provider and the recipient, use of personal self-care practices, creating healing relationships, applying the knowledge of health promotion and maintenance, and the appropriate integration of complementary and conventional medicine practices. Nelson et al. [7] described "healing environment" as synonymous with the therapeutic environment. The therapeutic environment is one that is designed to not only support and facilitate state-of-the-art medicine and technology, patient safety, and quality patient care, but also to embrace the patient, family and care providers in a psycho-socially therapeutic environment. From the above, it could be concluded that the therapeutic or healing environment is a part of the spiritual environment of the Thai elderly. The concept of spiritual well-being and healing environment would be checked and compared between the mentioned literature and existing cases in Thailand. The conclusion would then be used in the questionnaire design.

Burgstahler [8] has defined universal design as designing any product or environment that involves the consideration of many factors, including aesthetics, engineering options, environmental issues, safety concerns, industry standards, and cost. Designers often focus on the average user [8]. The National Disability Authority [9] has defined universal design as the design and composition of an environment so that it can be accessed, understood and used to the greatest extent possible by all people, regardless of their age, size, ability or disability. The Center for Universal Design [10], established seven principles of universal design to provide guidance in the design of products and environments. The principles include:

- Equitable use - the design is useful and marketable to those with diverse abilities.
- Flexibility in use - the design accommodates a wide range of individual preferences and abilities.
- Simplicity and intuition - use of the design is easy to understand, regardless of the user's experience, knowledge, language skills, or current concentration level.

- Perceptible information - the design communicates necessary information effectively to the user, regardless of ambient conditions or the user's sensory abilities.
- Tolerance for error - the design minimizes hazards and the adverse consequences of accidental or unintended actions.
- Low physical effort - the design can be used efficiently, comfortably, and with a minimum of fatigue.
- Size and space for approach and use - appropriate size and space is provided for approach, reach, manipulation, and use, regardless of body size, posture, or mobility.

Paul et al. [11] stated that a community is a social unit of any size that shares common values, or is situated in a given geographical area (e.g. a village or town). It is a group of people who are connected by durable relations that extend beyond immediate genealogical ties, and who mutually define that relationship as important to their social identity and practice. The WHO Regional Office for Europe [12] defined "a community residential health facility as a non-hospital, community-based mental health facility that provides overnight residence for people with mental disorders". The facilities include: supervised housing unstaffed group homes, group homes with some residential or visiting staff, hotels with day and night staff, hostels and homes with 24-hour nursing care staff, halfway houses, and therapeutic communities. Both public and private not-for-profit and for-profit facilities are included.

Perkins et al. [13] stated in a textbook of "building type basics for senior living" that common facilities within a skilled-nursing facility serving all of the nursing units may include: multipurpose room, coffee shop/snack bar, gift shop, library, outdoor terraces, recreation areas, art/activity, clinic, rehabilitation. The textbook also stated that, in adult communities, landscaped and natural areas should be developed for walking, contemplation, golf, lawn, sports, shuffleboard, gardening, fishing, and other recreational activities.

Jabareen [14] defined a conceptual framework as a network or plane of interlinked concepts that, together, provide a comprehensive understanding of a phenomenon or phenomena. Each concept of a conceptual framework plays an ontological or epistemological role in the framework. Jabareen [14] also suggested 8 phases in the procedure of conceptual framework analysis including: (1) mapping the selected data sources, (2) extensive reading and categorizing of the selected data, (3) identifying and naming concepts, (4) deconstructing and categorizing the concepts, (5) integrating concepts, (6) synthesis, re-synthesis, and making it all make sense, (7) validating the conceptual framework, and (8) rethinking the conceptual framework.

3. Research methods

The Delphi technique was developed by Olaf Helmer and Norman Dalkey in the 1950s and is still a widely used and accepted method for achieving convergence of opinion concerning real-world knowledge solicited from experts within certain topic areas [15]. The technique is designed as an expert group communication process and aims to conduct detailed examinations and discussions of a specific issue. The process is continuously iterated until a consensus is determined to have been achieved. This research used the Delphi technique to confirm the factors and framework of environmental residential communities to enhance the well-being of the elderly in Thailand. Steps of the Delphi technique were as follows:

Step 1. Identify the factors: Informal interviews and observations were conducted for three case studies in Thailand in terms of management processes and environmental designs. Interview results were summarized along with related theories and literature such as textbooks research articles (ISI, SCOPUS and ScienceDdirect databases) and related annual reports in Thailand. All factors were collated and listed before proceeding to the next step.

Step 2. Confirm the factors: The listed factors were confirmed by seven experts in related fields using the Delphi technique. The experts were selected based on their professional experience as follows:

- Three experts involved in the management of residential communities for the elderly, working in both government and private sectors.

- Two experts in elderly studies from academic institutions.
- Two experts in design aspects of residential communities employed by private companies.

The listed factors were confirmed by a consensus of the seven experts' opinions directly and individually starting from expert 1, 2, 3..., 7 and returning to expert 1 circularly. All experts carefully and individually considered the listed factors. Anonymous responses were aggregated and shared with the group after each round until a consensus was saturated. Once the listed factors were confirmed and categorized by the seven experts, they were constructed as the conceptual framework in the next section.
Step 3. Draft the conceptual framework: According to the expert opinions, confirmed factors were categorized and arrows were drawn between the factors as a draft conceptual framework. This was then reconfirmed by the seven experts by repeating Step 2. The final framework was considered as the conceptual research framework.

4. Results

Opinions of the seven experts as confirmed factors, items as indicators of the Thai elderly's requirements in healing environment and universal design of residential communities are listed in table 1.

Table 1. Factors for Thai elderly's requirements and their measurement items.

Factor		Item
Elderly's Characteristics		CH1: Age CH2: Health
Elderly's Requirement	Healing Environment	Q1: Liking natural light during the daytime
		Q2: Liking the sounds of nature such as birds singing, water flow
		Q3: Liking listening to music during the day
		Q4: Liking the smell of flowers in the residential area
		Q5: Focusing on "colors" (light-dark colors, bright-pensive colors) in the residence
		Q6: Giving importance to "shapes" (geometry, natural and independent shapes) in the residence
		Q7: Liking the paintings on the walls
		Q8: Liking the sculptures displayed on tables or floors
		Q9: Liking flowers in vases for decoration of the residence
		Q10:Attention to the texture (soft, hard, rough, smooth) of the floor, wall, furniture in the residence
	Universal Design	Q11: Need for a buoyancy aid (handrails) for walking
		Q12: Need for a buoyancy aid (handrails) for sitting-up
		Q13: Need for a flat surface floor of the residence (no stairs)
		Q14: Need for ramps instead of stairs
		Q15: Liking a bright environment
		Q16: Liking an orderly environment
		Q17: Need for road crossing signals in the residential community
		Q18: Need for sound signals when there is an emergency in the residential community
		Q19: Need for a transmitter to receive a signal in the residence when there is an emergency
Willingness		W1: Interested in the residential community W2: Willingness to live in the residential community

Table 1. Factors for Thai elderly's requirements and their measurement items.

Factor	Item
	W3: Willingness to recommend the residential community to others

In table 1, factors of the elderly's characteristics comprised age and health as CH1 and CH2. The elderly's requirements comprised two sub-factors as (1) healing environment; 10 items (Q1 to Q10), and (2) universal design; 9 items (Q11 to Q19), while willingness comprised 3 items (W1 to W3).
All factors for Thai elderly's requirements and their measurement items were used to construct the conceptual framework shown as figure 1. This framework was extended to include more detail of the items, variables and their indicators as shown in figure 2.

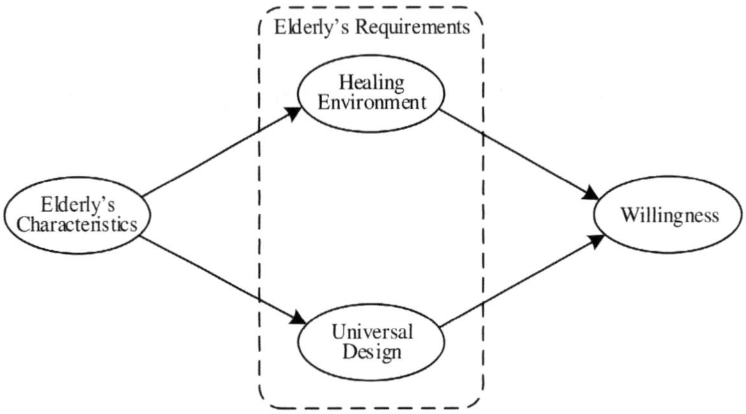

Figure 1. Conceptual framework of residential design to enhance the well-being of the elderly in Thailand.

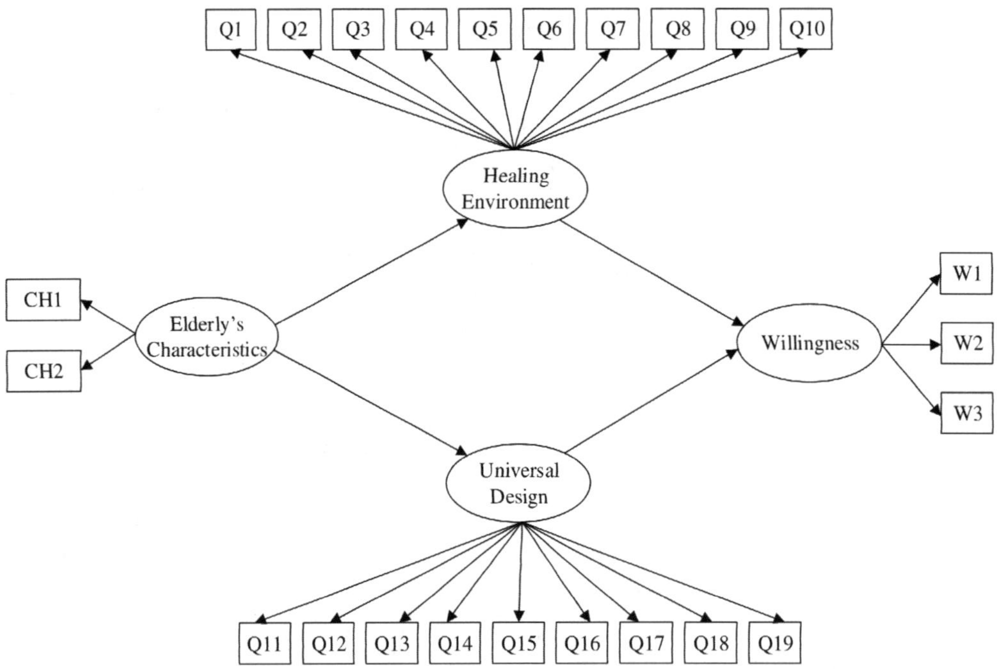

Figure 2. Detailed conceptual framework.

5. Conclusions

This research studied factors related to the Thai elderly's requirements to enhance their well-being in a residential community by modeling a conceptual framework. Three case studies in Thailand were observed and interviews were conducted in terms of environmental design and management. Factors were identified from the three case studies and confirmed by seven experts using the Delphi technique. Results indicated that requirements of the elderly in a residential community included age and health as CH1 and CH2, healing environment (Q1-Q10), universal design (Q11-Q19), and three factors of willingness (W1-W3) as shown in table 1. The conceptual framework, as the outcome from the Delphi technique (figure 2), showed that factors of the elderly's characteristics had a positive direct effect on their requirements in a residential community. These requirements comprised two sub-factors including (1) healing environment, and (2) universal design. All factors impacted on the willingness of the elderly to join communal events of the residential community. The framework in figure 1 was extended to include more details of the items, variables and indicators as shown in figure 2. This framework will be used for quantitative analysis to determine future policy for environmental design and management of residential communities to enhance the well-being of the elderly in Thailand.

6. References

[1] The Thailand Nation Elderly Committee, *The Yearly Report of Thai Elderly Situation 2012*,The Ministry of Social Development and Human Security Thailand, Bangkok, Thailand, (2012)

[2] The National Committee for the Elderly, the Ministry of Social Development and Human Security Thailand, *The 2nd National Plan for the Elderly (2002-2021)*, The Ministry of Social Development and Human Security, Bangkok, Thailand, (2009)

[3] Information on http://www.who.int/healthinfo/survey/ageingdefnolder/en/

[4] R. Plunz, *Design and the Public Good,* Massachusetts Institute of Technology. Cambridge, MA. USA, (1982)

[5] World Health Organization, *Constitution of the World Health Organization (Article 80) (Geneva: WHO)*(1946)

[6] W.B. Jonas and R.A. Chez: Toward Optimal Healing Environments in Health Care: submitted to The Journal of Alternative and Complementary Medicine Vol. 10 Supplement 1 (2004)p. S-1–S-6

[7] C. Nelson, T. West and C. Goodman: The hospital built environment: what role might funders of health services research play? (Rockville MD: Agency for Healthcare Research and Quality 2005 Aug Contract No 290-04-0011 AHRQ Publication No 06-0106-EF) (2005)

[8] S. Burgstahler *Universal Design: Process, Principles, and Applications,* University of Washington, College of Engineering, UW Information Technology, College of Education, Washington(2015)

[9] Information on http://universaldesign.ie/What-is-Universal-Design/The-10-things-to-know-about-UD

[10] Information onhttps://www.ncsu.edu/ncsu/design/cud/about_ud/udprinciples.htm

[11] J. Paul, Nadarajah, Yaso, H. Karen, and S. Victoria, *Sustainable Communities*, Sustainable Development: University of Hawaii Press, Other Paths for Papua New Guinea, Honolulu, (2012), p. 14

[12] WHO Regional Office for Europe, *Policies and Practices for Mental Health in Europe*, Copenhagen, Denmark, (2008), p. 71

[13] B. Perkins, J.D. Hoglund, D. King, and E. Cohen, *Building Type Basics for Senior Living*, John Wiley & Sons, Inc., New Jersey, USA, (2004)

[14] Y. Jabareen, Building a Conceptual Framework: Philosophy, Definitions, and Procedure: International Journal of Qualitative Methods, 8(4) (2009), p. 49-62

[15] Hsu, Chia-Chien, and B.A. Sandford: The Delphi Technique: Making Sense of Consensus:Practical Assessment, Research & Evaluation, Vol. 12, No. 10, (2007)

IOP Conf. Series: Materials Science and Engineering **829** (2020) 012010 doi:10.1088/1757-899X/829/1/012010

Acknowledgments
This research project was financially supported by Mahasarakham University, Thailand.

IOP Conf. Series: Materials Science and Engineering **829** (2020) 012011 doi:10.1088/1757-899X/829/1/012011

Evaluation of Acoustic Environment of High-Rise Residential Space

Yumeng Cheng

School of Architecture, Harbin Institute of Technology; Key Laboratory of Cold Region Urban and Rural Human Settlement Environment Science and Technology, Ministry of Industry and Information Technology, Harbin,150001,China

1197861493@qq.com

Abstract. As the most common type of urban residential form, high-rise residential space is the main place for people's leisure and entertainment, so its acoustic environment quality is very important. In this paper, the acoustic environment of typical high-rise residential space is measured and analyzed, and survey questionnaires are statistically analyzed to explore the influence of listener's behavioural factors and social factors on the perception of soundscape. In order to provide reference for related design and management work. The results show that behavior factor such as activity frequency, time, location and social factors such as sex and age was associated with soundscape evaluation significantly.

1. Introduction

Architectural design is not only the creation of space forms, but also the shaping of the overall environment. At present, most studies about residential space's acoustic environment only focus on the impact of traffic noise [1, 2]. However, traffic noise is not the only sound source that can disturb people. Especially in high-rise residential space, the population can be doubled, corresponding the sound types and their effects will be more complex. Previous studies have shown a lack of attention to the effects of living sound sources. In additions, some studies have shown that people's perception of sound is affected by factors such as the type of sound source, volume, and listeners's individual differences, but not only sound pressure level (SPL) [3, 4]. Therefore, based on the analysis of sound field in high-rise residential space, this paper focuses on the sound source, to explore what sound sources that can really affect residents, how do people evaluate these sound sources, and what factors influence the subjective evaluation.

2. Methods

2.1. Site selection

This paper chooses the European new town community in Harbin as the investigation object, mainly based on three reasons. First, Harbin is a typical city in cold region in China, the provincial capital with the highest latitude and lowest temperature. In cold regions, the climate is cold in winter and lasts for a long time, In order to resist the attack of cold wind, the residential buildings are mostly enclosed and centralized. The enclosed building form of this residential space can representative cold region . Secondly, the residential occupancy is high and the age distribution of residents is balance, which is convenient for sampling survey. Thirdly, there are abundant sound sources in the community, and each active spaces have different functions, covering different sound sources, so the sound field features are significant. Therefore, this residential space is selected as the research object.

2.2. Questionnaire survey
In order to get an objective and effective evaluation, residents were randomly selected for the questionnaire survey. Because of the small number of outdoor activities in Harbin in winter, the research was conducted in summer. The questionnaire questions are divided into two parts: one part is the background information of each respondents: Including social factors such as gender, age, work, residence time and floor height, as well as behavioral factors such as frequency of activities, activity items, duration, site. The second part is acoustic environment evaluation: Includeing an assessment of the sound source's volume and preference, and then SPSS was used for data entry and correlation analysis.
Among the 220 valid questionnaires collected, the ratio of male to female was 4.5:5.5. In terms of age distribution, there are more elderly over 60 years old, accounting for 24.9%. Retirement accounted for the highest number in working condition, accounting for 36.1%. Most of the residents lived for less than 3 years, accounting for 34.4%. Most of the outdoor visitors are living in the upper floors, 20.9% of them live above the 15th floor, 18% of the 10-14 floors, this basically according with the rule that the higher the residential floor, the more inclined the residents are to outdoor activities. In the activity item, playing with children topped the list at 26.5 percent.

2.3. Measurement method
At the same time of questionire, BSWA801 sound level meter was used to record the sound pressure level(SPL). The tests were conducted on weekdays, The sound level meter placed horizontally during measurement, the probe was 1.2m away from the ground, 1m away from the external wall , and 3.5m away from the reflector . The data were recorded once every 10s and tested continuously for 5 minutes at each point for a total of 30 times. The arrangement of measurement points is shown in the figure 1.

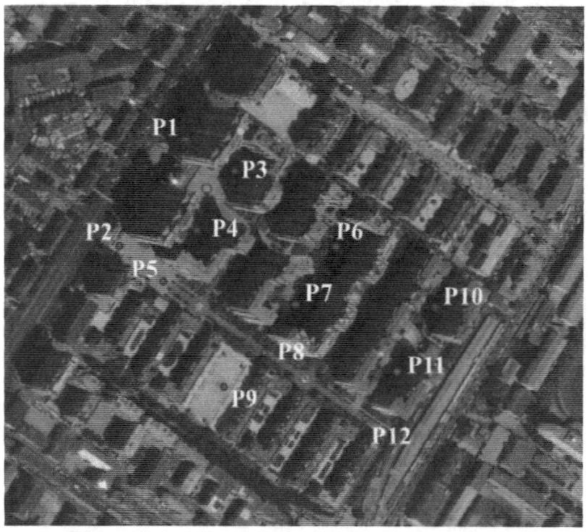

Figure 1. Top view of test point area.

3. Characteristics and evaluation of acoustic environment

3.1. Characteristics of the acoustic environment
The distribution of sound is closely related to the spatial function. Each space has different functions and activity , contains different sound sources, and also presents different sound pressure level(SPL) in different times. According to the description of sound sources given by the interviewees in the survey, the sound sources can be roughly divided into natural sound, animal sound, artificial sound, musical sound, mechanical sound and traffic sound.
At 13:30 pm, it is the rush hour, traffic sound is the main sound source. As shown in figure 2, the SPL

of road space (P1 and P2) was up to 68.9dB, which was significantly higher than that of other measuring points. At this time, the number of people in the courtyard (P3 and P4) was relatively small, acoustic environment was mainly influenced by the sound of traffic. Due to the shielding of buildings, the SPL in this courtyard was 10.7dB lower than road. Because the residential's internal road (P5, P8 ,P12) have no shelter to the direct traffic sound, so its SPL is 5.3dB higher than adjacent yard .

During the working time of 15:30 pm, SPL at all points changed significantly. They were all higher than noon, and the sound of propaganda trumpets and weeding machines were the main noise. There was a publicity campaign at P5, therefor the SPL increased to 11.9dB due to the trumpets sound, reaching the highest SPL of 75.4dB in the whole day. At P10 and P11, there were a weeding machines working, so the SPL increased by 21.8dB. Except for P5, the SPL of other points which outside the yard was lower than courtyard, which indicated that the influence of living noise in this period was serious.

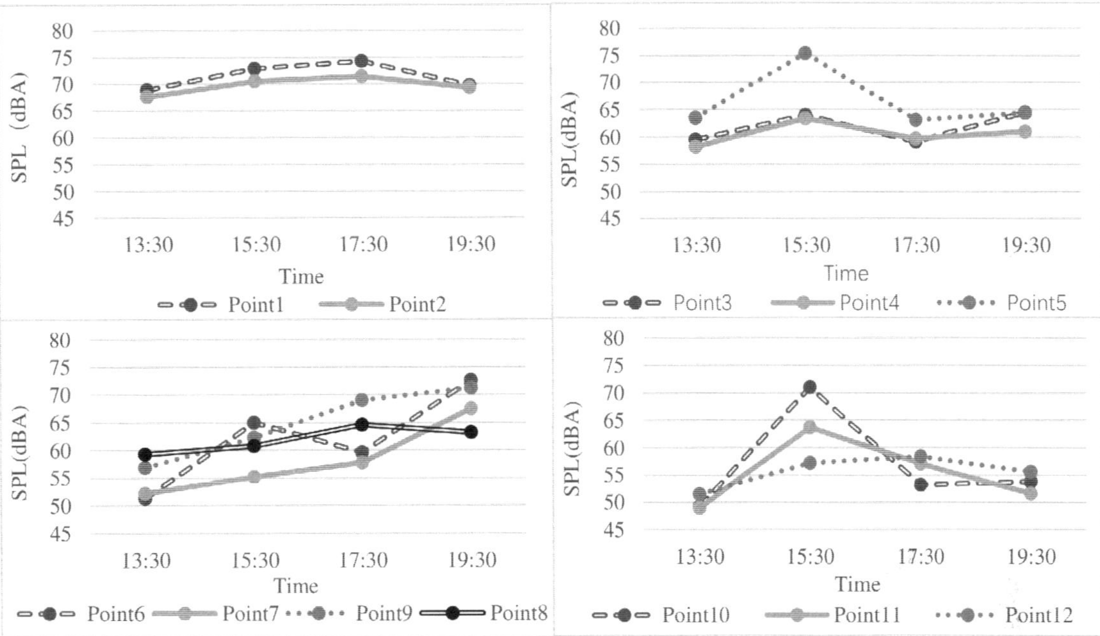

Figure 2. Sound pressure level (SPL) of each measurement point.

At 17:30 pm, it is the rush hour, traffic noise is the main source, and the peak of SPL is still at the P1 and P2, reaching the peak traffic noise(74.3 dB) of a day. During this period, crowds began to gather in the east square (P9), and the main sound sources were chatting sound, children's frolicking sound and musical instrument sound. The SPL was increased by 6.7dB.

At 19:30 pm, it is the leisure time for residents after dinner. The number of residents in courtyard reached the highest in a day, and life noise was the main sound source. The peak of SPL at P6 (72.6dB), which was higher than the traffic noise by 2-3dB. There has public fitness facilities in this yard, so the main sound sources are children's frolicking, chatting, dogs barking and musical instruments. At the P9 (71.2dB) has square dance activity, so the music and square dancing sound were very noisy. Furthermore, P10 and P11 were in the innermost courtyard, with no sports equipment and the highest greening rate. The main sound sources were birdcalls, leaves rustal sound, and the courtyard had the lowest SPL (51.6dB).

3.2. Evaluation of sound sources

In terms of sound source preference evaluation, the questionnaire results (figure 3) showed that residents' favorite sounds were leaves rustle sound 63%, birdcalls 71.6%, and music 42.1% .

In the artificial voice, although chatting's volume is moderate, but it's evaluation uncomfortable. In contrast, although the sound of children's frolicking and square dancing was louder, but good evaluation. There have 47.6% and 34.2% of the residents like children's frolicking sound and square dancing sound, while only 14.5% are bored. This is not consistent with previous conclusions and expected low rating[5]. Combined with the survey results of residents' activities, it can be found that there are most elderly respondents in this community, and their main activities are playing with children and dancing in the square. Therefore, they are more receptive to such sounds. By contrast, chatting is hard to engage in, people always get annoyed with such sounds that have nothing to do with them.

The most annoying sounds for residents are mechanical sound such as propaganda trumpets 87.1%, construction noise 82.4%, weeding machines' sound 69.3% and electric cars' sound 49.4%. As well as traffic noise such as cars horn 73% and traffic passing sound 72%. Combined with objective data measurement, it can be seen that these sound always have high SPL and strong interference to surrounding residents.

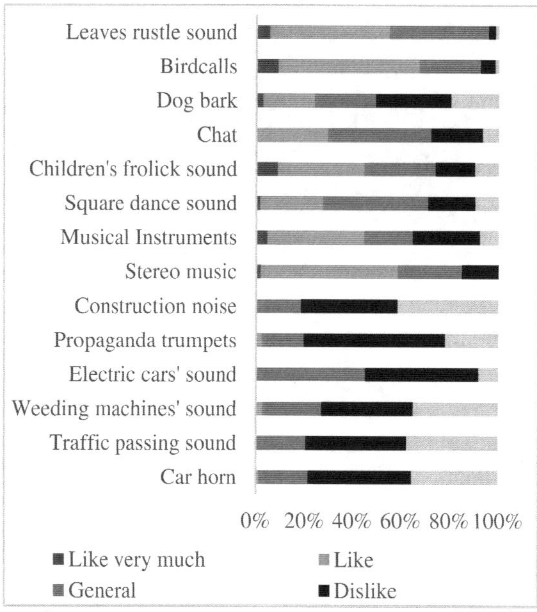

Figure 3. Acoustic preference evaluation.

4. Influencing factors of acoustic environmental assessment

4.1. Behavioral factors

4.1.1. Frequency and time. The relationship between residents' behavioral factors and acoustic environment evaluation is shown in the table 1. It was found that the visit frequency of residents was significantly correlated with the evaluation of traffic sound at $p \leq 0.01$. Indicates that the more often the residents activities in the yard, the greater the tolerance to traffic noise. Residents' duration time is significantly correlated with children's frolicking sound, and comfort evaluation of acoustic environment and green environment. It indicates that the longer people stay in the residential space, the more acceptable to the noise of children, and the evaluation of acoustic environment and green environment is better, which is consistent with the previous research conclusions [6].

Table 1. The relationship between behavioral factors and soundscape evaluation.

Sound source	Activity frequency	Activity project	duration time	site
Dog bark	0.054	0.051	0.149	-0.361**
Chat	0.015	0.024	-0.013	-0.243**

Children's frolick	-0.147	0.033	-0.235**	-0.070
Musical Instruments	-0.047	0.008	0.046	-0.280**
Construction noise	0.121	0.079	0.054	-0.231**
Electric cars	-0.175	-0.020	-0.039	-0.419**
Traffic pass	-0.249**	-0.122	-0.128	0.355**
Car horns	-0.204**	-0.163	-0.153	0.310**
Acoustic environment	-0.059	0.002	-0.172*	0.179*
Green environment	0.094	0.086	-0.177*	0.10

Note: * and ** indicate the significance level, * means $p \leq 0.05$, and ** means $p \leq 0.01$

4.1.2. Activity site. The activity site of the listeners was significantly positively correlated with the traffic noise and the acoustic environment at $p \leqslant 0.01$ and $p \leqslant 0.05$. This indicates that the closer the site is to the street, the lower evaluation of traffic noise, and also the lower evaluation of acoustic environment. On the contrary, activity site was significantly negatively correlated with dog barking, chatting, musical instruments, construction and electric cars. while these sound sources were all living noise, indicating that the farther away from the road, the lower the evaluation of living noise. This results are consistent with objective test's data.

4.2. Social factors

*4.2.1. Gender factors.*The respondents' gender only affected some sound sources (table 2). Gender was positively correlated with birdcall at $p \leq 0.05$, and negatively correlated with dog barking and construction sound. Combined with the acoustic preference evaluation (figure 3), it can be seen that the acoustic preference value of birdcall is higher, while dog barking and construction sound is lower. Presumably in high-rise residential space, compared with men, women feel more strongly to sound like or dislike.

4.2.2. Age factor. Previous studies on urban soundscape have found that age is correlated with acoustic comfort [7, 8]. However, in this study, age was only correlated with dog barking evaluation at the level of $p \leqslant 0.01$. This suggests that older listeners in higher residential areas are more sensitive to animal sounds, such as the dog barking sound.

Table 2. The relationship between social characteristics and sound source preference.

Sound source	Gender	Age	Job	Residence time	Floor height
Birdcalls	0.196*	-0.127	0.166	0.161	0.161
Dog Bark	-0.218*	-0.281**	0.059	0.069	0.069
Construction noise	-0.388**	-0.032	0.186	0.211	0.211

Note: * and ** indicate the significance level, * means $p \leq 0.05$, and ** means $p \leq 0.01$

5. Conclusions

Based on the study of sound field and acoustic environment evaluation in high-rise residential space, the following conclusions can be drawn:

1. In the high-rise residential spce, people have the highest evaluation on the natural sound. Different from previous research conclusions, the loudness and preference of children's frolicking sound and square dancing sound are both higher.
2. Behavioral factors influence acoustic environment assessment. The higher the frequency of visits, the stronger the listener's tolerance to the traffic noise. The longer residents stay in the residential space, the higher the evaluation of children's frolicking sound, acoustic environment and green environment.
3. Social factors influence acoustic environment assessment. Compared to men, women feel more

strongly to sound. Different from previous studies, age only related to animal's evaluation, the older the listener is, the more sensitive he or she is to animal sounds such as the dog barking.
In this study, it was found that in addition to the influence of sound pressure level and listener's personal differences, the acoustic environment of high-rise residential space was also affected by architectural layout. I will make further discussion in the next research.

6. Reference

[1] Sanchez G M E, Renterghem T V and Botteldooren D 2016 The effect of street canyon design on traffic noise exposure along roads *Building and Environment* **97** 96-110
[2] Zhou,Z Y, Jin H and Kang J 2011 Study on the simulation and optimization design of the building form along the street under the influence of traffic noise *Architecture science* **21(3)** 222-224
[3] Yang W and Kang J 2005 Acoustic comfort evaluation in urban open public spaces *Applied Acoustics* **66** 211-229
[4] Meng Q, Kang J and Jin H 2010 Research on the influence of social factors on the subjective loudness and harmony comfort of underground commercial street *Application of acoustic* **29(05)** 371-381
[5] Ren X X, Kang J and Liu X G 2016 Study on the perception of urban leisure green space*Landscape architecture* **04** 42-55
[6] Meng Q, Jin H and Kang J 2010 Influence of user behavior on subjective loudness and harmony comfort of underground commercial street *Huazhong architecture* **05** 90-92
[7] Kang J 2004 *Urban Sound Environment* (UK:Taylor & Francis) p73-76
[8] Nilsson M E and Berglund B 2006 Soundscape quality in surburban green areas and city Parks *Acta Acustica united with Acustica* **92** 903-911

The Same-floor Drainage for Independent Retrofitting of the Existing Apartment Buildings without S-I System in China

ZHU Ning [a]; JIANG Yong [a]; WANG Qiang [b]; GUAN Wenmin [c]

a. School of Architecture, Tsinghua University, Beijing, 100084, China
b. Student Community Management & Service Centre, Tsinghua University, Beijing, 100084, China
c. SPEC Plumbing Co. Ltd., Ningbo, 315514, China

[a] 13717742900@163.com

Abstract. By 1998 the total floor area of urban residential buildings was 9.59 billion square meters in China. In this period, most of the apartment buildings in China adopted brick-concrete structure or shear wall structure without SI-system, and limited the integrating means of retrofitting and reuse. Because of the strict control of area standard in urban residential construction at that time, the toilet area is always too small to carry the contemporary life style, which takes the first difficulty within independent retrofitting of the apartment building. Since the same-floor drainage system is widely used many new built apartment buildings, this paper aims to research and develop this system into retrofitting of the existing apartment buildings without S-I system in China. Based on the analysis to the household plans and their retrofitting design, two kinds of same-floor drainage systems are employed within the toilets, solve the problems of pipe connection, floor elevation and waterproof, and succeed to improve the function and space use of whole household.

1. Introduction

China's large-scale urban apartment buildings construction began in the 1950s. By 1998 the total floor area of urban residential buildings was 9.59 billion square meters [1]. Among them, a large number of residential buildings adopt brick-concrete structure or steel-concrete shear wall as the main load-bearing structure (figure 1). Because of the strict control of area standard in urban residential construction at that time, it was unable to carry the contemporary life style [2]. As its high similarity of household type, there are much greater potential values within retrofitting in the household scale.

Figure 1. Sorts of elevation of the apartment buildings in inner Beijing area

1.1. Non-SI system characteristics of existing apartment buildings in China

The basic theory of renovation of apartment buildings came from Skeleton-Infill system proposed by Professor Habraken of the Netherlands in 1961 to alleviate the contradiction between large-scale housing construction and diversified demand of living style [3]. After the 1970s, the Japanese Residential Communities developed their own KSI residential system by using the SI system of the

Netherlands for reference, so as to facilitate the renovation and maintenance of buildings, focusing on the renewal and reuse of products from the community to the residential unit to the interior of the house [4]. In this period, most of the apartment buildings in China adopted brick-concrete structure or shear wall structure (figure 2).

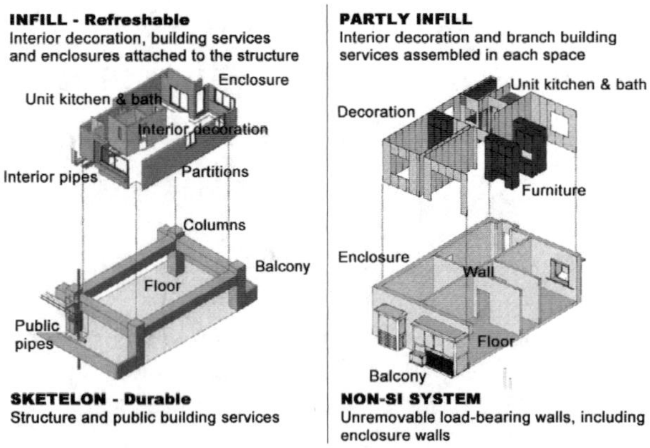

Figure 2. Compared with S-I system (left), space inflexibility in the load-bearing walls system limited the integrating means of retrofitting and reuse (right).

1.2. Main problems and difficulties in reform of toilet in apartment buildings

In these types of apartment buildings, the toilet area is always too small to contain the enough numbers of sanitary wares, and even the number of drainage outlets is insufficient. Taking 1974's apartment building style as an example (figure 3), the net area of its toilet is only 1.36 square meters. The original design includes 1 squatting toilet, 1 water tap for rinsing, and 1 floor drain. There is no shower space, and even no wash basin. During more than 40 years, wash basin and shower spraying were gradually added to the bathroom, and some toilets were replaced, but still inconvenience to everyday life.

Figure 3. Early squatting toilets using water outlet and cast iron pipes, required a large backwater bend to ensure the water seal of drainage branch pipes, using the roof part through the floor in the lower households to install and access drainage risers. In order to ensure the bedding and pedal of the squatting toilet, it has to raise the floor level around the toilet by 200-400 mm.

1.3. The same-floor drainage system of the new-built apartment buildings in China

The same-floor drainage is a special technology for connecting drainage pipes to drainage risers on the same floor as the household space. The advantages of the same-floor drainage are as follows [5]:

- Clear property rights of household
- Waterproof safety with small leakage probability
- Flexible change of layout of sanitary appliances
- Low drainage noise
- No condensation underwater drip

1.4. The research method (figure 4)

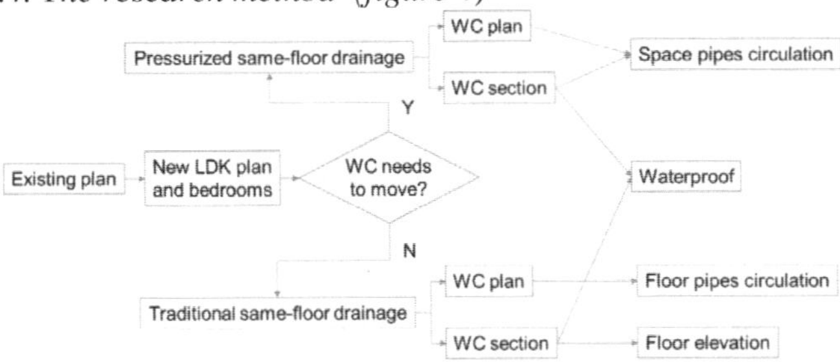

Figure 4. Routine of design and development to the same-floor drainage system in existing apartment buildings. It depends on the LDK integration in the plan as well as the number of bedrooms.

2. Traditional same-floor drainage retrofitting: renovation in the expanding toilet area

2.1. Family characteristics

At most three generations and six people live together: two children under school age, two couples of young teachers, and one or two old people or nannies who take care of the children. Old people usually concentrate on activities in the kitchen and dining room, while keeping an eye on children in the living room. Therefore, a set of visual LDK space is very important. Considering the situation of large family population, as early rise, or bedtime, it is also very important for each sanitary ware to have independent use space (figure 5).

2.2. Technical difficulties and opportunity points

Because of the non-SI system, the original bathroom area is too narrow to carry out three-apart separation design. In addition, the same-floor drainage needs to raise the elevation by 80-120 mm, and lay the horizontal branch pipe with the diameter of 50 mm leaking to the plumbing riser. If all other rooms are raised with the bathroom, it means that not only compresses the net height of the use space, but also a lot of construction costs. Carefully studied the characteristics of the original house space, the above two problems can be solved in three favourable opportunities.

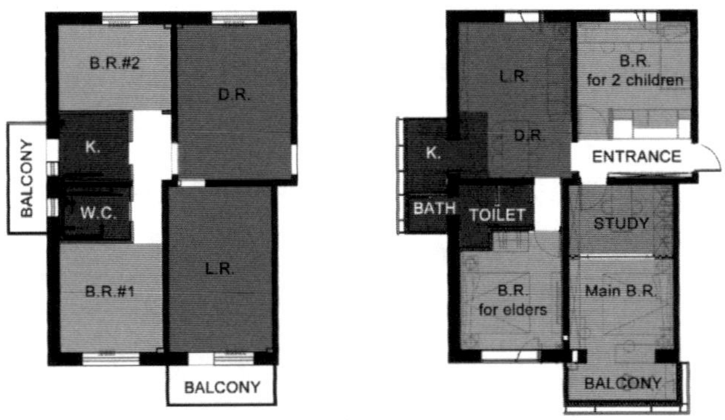

Figure 5. Opportunity points: a) LDK interconnection ensures the area of basic activity space. b) As the window was turned into a door, the space of the toilet expands to the balcony as a bath room. c) Kitchen can also move to the balcony and give space to dining room.

2.3. Renovation design and development

According to the requirement of the same-floor drainage technology system, the design of drainage pipeline connection, surface elevation and waterproofing of floor are mainly considered.

2.3.1. Drainage pipeline connection. There are four drainage points in the toilet (figure 6). The advantage of the same-floor drainage is that it can redesign the location of each sanitary ware without relying on the original toilet layout and drainage point, so as to make its layout more reasonable. Reform of the existing main plumbing pipe without disturb to neighbors above and under needs to employ hydrocyclone as the collection of branches of sanitary, while plugging into the root of existing pipe and connecting the branches pipes above (figure 7).

IOP Conf. Series: Materials Science and Engineering **829** (2020) 012012 doi:10.1088/1757-899X/829/1/012012

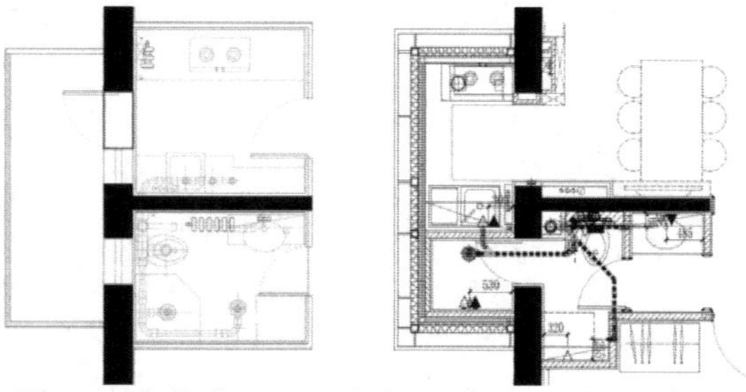

Figure 6. Before (top left) and after (top right) retrofitting of the plumbing system.

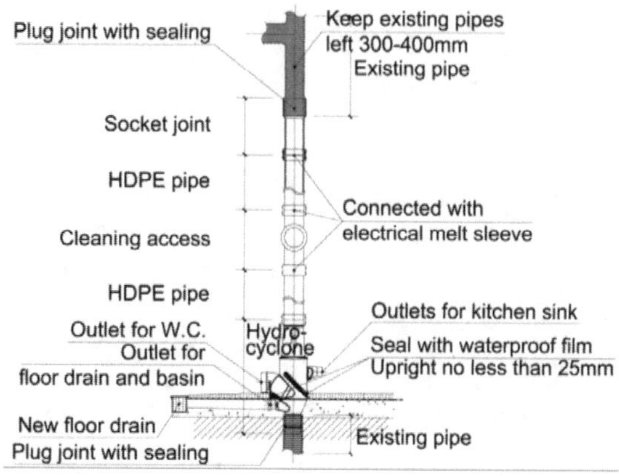

Figure 7. Detail design of changing main pipe of the plumbing with hydrocyclone.

2.3.2. Floor elevation. Elevation design is the key to the design of the same-floor drainage, which determines the final completion of floor surface. On the one hand, it is necessary to ensure that the gradient of drainage is not less than 1.2% [6] (the washing machine has pressure drainage, without considering gradient), on the other hand, it is necessary to minimize the completed elevation for more net height. In order to make the 50 mm drainage outlet of the cyclone strive for lower drainage point, the concrete cushion of 30 mm downward is also cut out, in order to ensure accessibility, the height of the floor drain in the bathroom was raised slightly twice in three spaces, each time 15 mm [7], to meet the requirements of the floor drain elevation in the bathroom. It is worth mentioning that the location of the height difference, combined with the traditional method of crossing the gate stone, makes the height difference difficult to detect, and the crossing gate stone has a slight chamfer transition (figure 8).

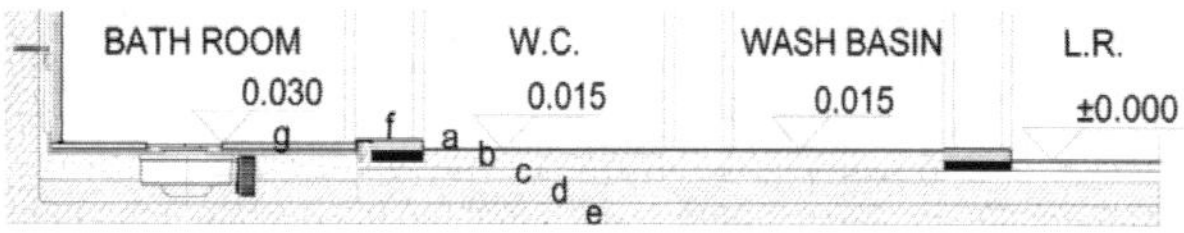

a) 5mm wood laminated floor panel; b) Self-leveling concrete; c) FR-Cement slab; d) 50mm existing cushion; e) RC floor; f) Gate stone; g) PP waterproof

Figure 8. Section of a series of toilet space, showing the detail of floor levelling.

2.3.3. Floor waterproof. Since the existing structural floor, including the balcony, employed a whole RC panel, there is no need additional membrane waterproof under or above the cement cushion layer. Only bathroom should it add whole unit waterproof to keep water penetrating the floor to the lower layers.

2.4. Retrofitting achievements (figure 9)

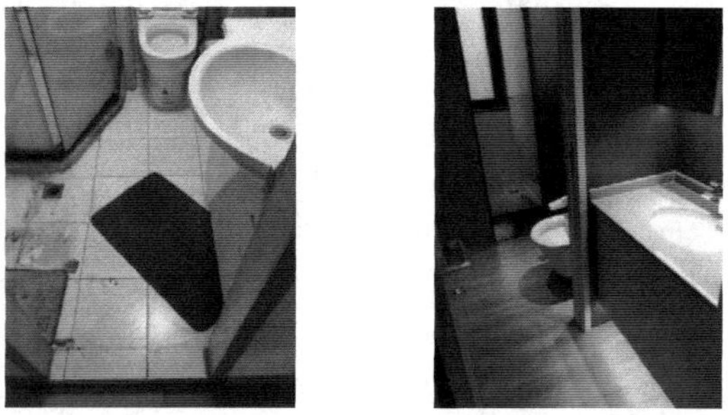

Figure 9. The retrofitted bathroom completely changed the original layout of sanitary ware by using the same-floor of drainage and according to the new way of life.

3. Pressurized same-floor drainage retrofitting: renovation of the whole displacement of toilet

3.1. Family characteristics

The family style is single or married residents without child (figure 10). Although the area of less than 40 square meters is not large, there is relatively abundant for everyday life, so comfortable, open space and quality of life are purpose of this retrofitting.

3.2. Technical difficulties and opportunity points
The living room cannot be directly illuminated. Near the entrance hall, the original small living space was more crowded. But the size of the bedroom, including a balcony, is too luxurious.

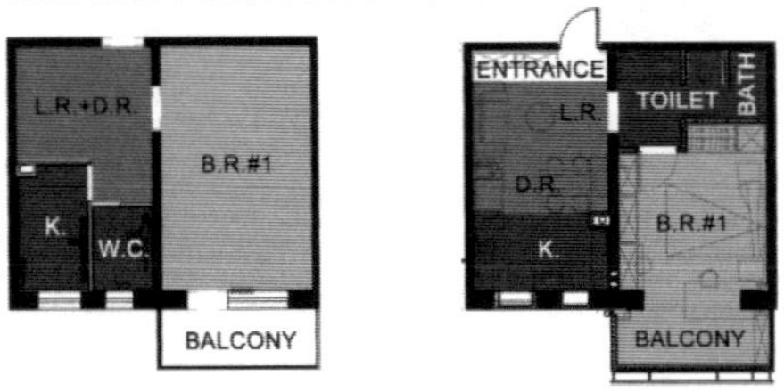

Figure 10. After the bathroom moves to the bedroom, the original living room, kitchen and toilet space can be fully integrated to form a typical large space of LDK. The entrance of the lobby can be dispersed in this space, which also solves the lighting problem of the living room.

The research team has come up with an idea. Since the same-floor drainage can make toilet redesigned, can it make the whole plan redesigned? There can be a try to move the bathroom to the bedroom.
It faces the following problem that the gradient of the same-floor drainage leads to elevation increase. The farther away from the public riser, the more elevation of the floor increases. In addition, moving the bathroom to the bedroom means that the bathroom on this floor stands above the bedroom on the lower floor. It is necessary to strengthen the waterproofing of the bathroom.

3.3. Renovation design and development

3.3.1. Spatial layout of pressurized same-floor drainage: Using sewage crushing and lifting pump, it is possible to reduce the diameter of drainage pipes for long distance plumbing, and at the same time to place them in the ceiling space, and share the service space with other pipelines. Finally, the parallel riser conveys downward to the riser through the hydrocyclone, so that the floor elevation can be leveled with the living room, as long as the elevation of the bathroom can meet the inlet water level of the lifting pump (figure 11).

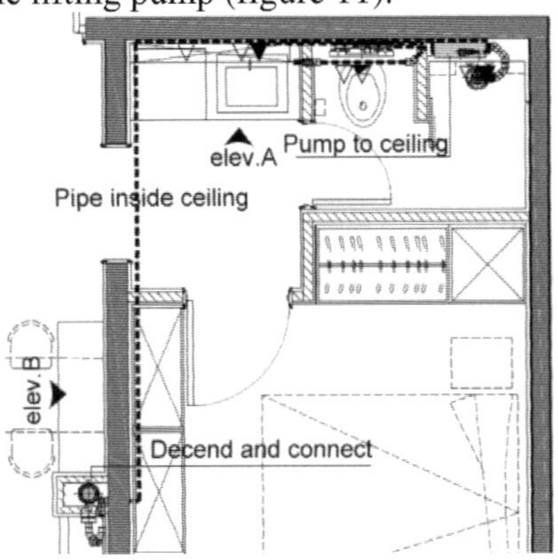

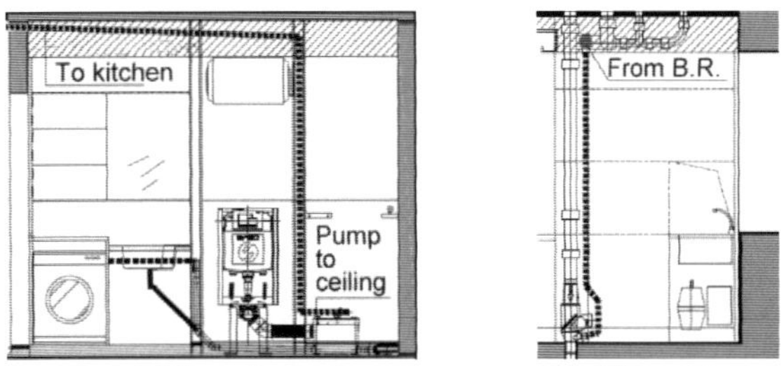

Elev. A Elev. B

Figure 11. Plan (left) and elevations of the pressurized same-floor drainage

3.3.2. Waterproof renovation of the original floor: Because the lower floor locates the bedroom, the same-floor drainage has to solve the problem that if the drainage pipeline would be damaged, it cannot penetrate directly to the lower floor, which avoiding the problem that the toilet cannot stand above other clean functional rooms in the building code. There will be also a violation of property rights of lower-class households, which originally should be the advantage of the same-floor drainage system. Thus, the bathroom employs two layers of waterproof to ensure not only the leakage of floor surface, but also the plumbing pipes (figure 12).

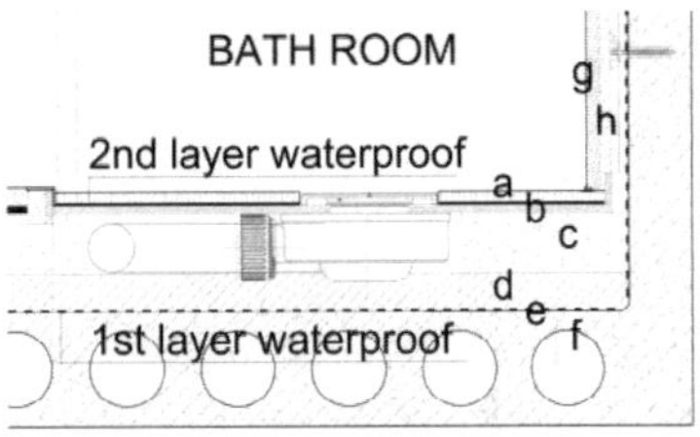

a) 10mm tiles with glue
b) 5mm PP waterproof
c) 100mm ceramsite concrete
d) 50mm cushion (new fill)
e) 2mm membrane waterproof
f) Prefabricated hollow panel
g) 20mm Alu-Polymer panel
h) 18mm calcium silicate panel

Figure 12. Detail of two layers of waterproof above the structural hollow floor panel, in contrast of the former case as the RC floor panel.

3.4. *Retrofitting Achievements (figure 13)*

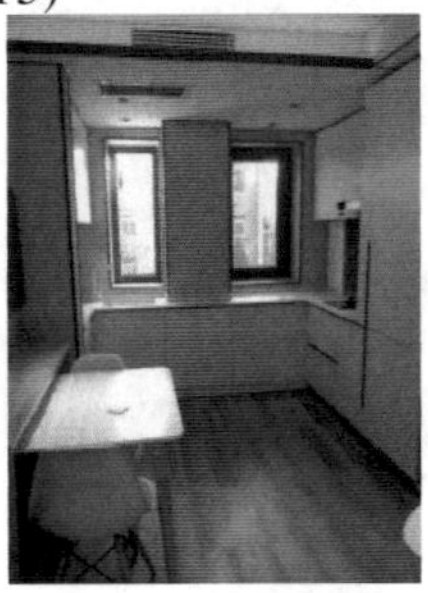

Figure 13. Through the overall move of the toilet space, LDK layout formed with good lighting conditions and space quality. At the same time, it improves the quality of bathroom, and does not disturb the everyday life of the upper and lower floor.

4. Discussions and prospects

4.1. Is there a contradiction between the overall movement of toilets and the relevant residential codes?
The original residential code requires that there should be no other type of room under the toilet [6]. In principle, if all drainage pipes are included in the overall space of the bathroom, then a whole closed waterproofing on the structural board can solve the problem of pipeline leakage, and if there is leakage, it will soon cross above the waterproofing weak parts such as the threshold, which will have a significant impact in the own household room, reminding to stop water and repair urgently. As the signal is obviously easy to detect, it will not disturb the household in lower floor.

4.2. Noise effect of sewage pump
It adopts the strategy to block the noise transmission path period by period. Firstly, the sewage lifting pump places in the bathroom, a corner formed by two load-bearing walls and structural floors, so that the running noise will not resonate with other light walls in the bathroom. Secondly, apart from the wall outside, the 600 mm thick wardrobe is partitioned between the bathroom and the bedroom, almost completely blocking the sound transmission from the partition wall. Then from the bathroom to the bedroom, there are three doors to block the air transmission more. Ultimately, when all doors are closed, it is not easy to detect the operation of the sewage pump in the bedroom.

4.3. The application prospect of the same-floor drainage in the household renovation of existing residential buildings
Using the opportunity of batch renovation of public housing household replacement period by the Housing Management Department of Tsinghua University, this research began to study the technical system of renovation through several model rooms from year of 2014. From 2015 to 2018, mass promotion was carried out according to the mature technology in the pilot project, and 117 households were renovated by the same-floor drainage system. Compared with the traditional downward drainage system, the advantages of the new technology system are obvious and easy to follow, especially in the existing apartment buildings without S-I system.

5. References
[1] Shao T 2013 General situation and development trend of urban housing supply in China *China Economic Times* **June 24** 005

[2] Lü JH, Rowe P G and Zhang J 2003 *Modern Housing in China 1840-2000* (Beijing: Tsinghua University Press)
[3] Matsumura S, Fan Y and Liu TT 2008 *Regeneration of residential area: New retrofitted apartment buildings in Europe and America* (Beijing: China Machine Press)
[4] Omi Y and Wu DH 2009 *Housing construction and industrialization in Japan* (Beijing: China Architecture and Building Press)
[5] Jiang Y, Guan WM, Zhu N and Chen DP 2018 *J. Human Settlements in Western China* **05** 28–33
[6] Ministry of Housing and Construction PRC 2009 *Code for Design of Building Water Supply and Drainage (GB 50015-2003)* **4.3.4** and **4.4.10** (Beijing: China Planning Press)
[7] Ministry of Housing and Construction PRC 2012 *Codes for Accessibility Design (GB 50763-2012)* **3.5.3** 7 (Beijing: China Planning Press)

Acknowledgments
This project is supported by the China National Key Research and Development Program (Project No. 2017YFC0702900) and the National Natural Science Foundation (Project No. 51978358 and 51578300). Thanks to the Housing Management Centre of Tsinghua University for providing a practical platform, and the technical support and integration are provided by the Research Centre for Sustainable Community, School of Architecture, Tsinghua University.

IOP Conf. Series: Materials Science and Engineering **829** (2020) 012013 doi:10.1088/1757-899X/829/1/012013

Local Climate Zones Definition in Relation to ENVI-met in the City of Dubai, UAE.

[1]Lindita Bande, [2]Prajowal Manandhar, [3]Prashanth Marpu and [4]Mohammad Al Battah,

United Arab Emirates University, Al Ain, United Arab Emirates
Khalifa University, Abu Dhabi, United Arab Emirates

[1]lindita.bande@uaeu.ac.ae

Abstract. The city of Dubai has experienced a rapid urban growth in the last 60 years. Located in the north west of the United Arab Emirates, Dubai has a linear development expanding mainly towards the coast. However recently this expansion is going towards the desert. Mapping such development creates a history that is relevant to many research entities. WUDAPT is a network that facilitates city mapping. With open source tools it is possible to create Local Climate Zones (LCZ). This zones have similar properties. Therefore, in the end of the process it would be possible to see the most dense urban areas of the city where the UHI phenomena is higher.
The process of mapping involves, Google Earth Pro, high definition satellite images and the CFD software, ENVI-met. The first two tools detect the LCZ and create the general map of the city. Then this map is compared with the validated models of ENVI-met. This comparison shows the connection of the UHI between one approach and the other. The site data taken from an open-source are used as a boundary condition in the simulations. The climate of Dubai is hot arid, with high temperatures in the summer and cool winters. The spring and autumn are short periods of 2-4 weeks.
The aim of this study is to detect problematic areas in the city in terms of surface/ air temperature, high levels of UHI. This detection helps understanding where should be intervened in order to have more walkable districts.

1. Introduction

The recent development of the cities around the world requires deeper analysis and evaluation. The expansion of the neighbourhoods, their characteristics need to be documented. The type of the materials applied, the green zones, the building distances and the general plan create the so called Local Climate Zones (LCZ). In several European Cities such as London, Milan, Rome, Amsterdam etch., the urban development is mapped and the information is available for the researches. This process help detecting the build-up area according to a template. It recognises the highest levels of the UHU phenomena. The link between landscape change and climate has been visualised in recent research on several urban cities, namely the World Urban Database and Portal Tool (WUDAPT). This is a global collaborative project dedicated to the acquisition, storage and dissemination of climate-relevant data on the physical geographies of cities worldwide. The city mapping has expanded in the Americas, India, China, etc. [1-3].
The second tool used in this study is ENVI-met. It's a CFD tool that simulates the wind speed, air temperature, relative humidity and much more parameters. The software in the version 4.0 has several

IOP Conf. Series: Materials Science and Engineering **829** (2020) 012013 doi:10.1088/1757-899X/829/1/012013

limitations such as the anthropogenic heat calculation. The tool still doesn't include it. The templates of the soil needs to be modified in order not to have overheated surfaces. However, ENVI-met is largely used in the urban planning area. It helps understanding and calculating the Outdoor Thermal Comfort as well. A relevant parameter especially for the middle eastern countries. In several other studies the focus in on the building itself. The latest version of the software permits detailed 3D models therefore more accurate results. The boundary conditions can be easily modified based on the available data of the specific study. In very detailed models (scale 250x250) the wind turbulences are visible in the plan and in the section, showing a clear image in different types of neighbourhoods. [4-6].

2. Methodology

The methodology shows the steps taken in this study. The city of Dubai was selected for this study considering the urban growth achieved in the last 50 years. The city has a hot arid climate with high levels of humidity, different from the cities inside the country such as Al Ain. [7-9]
The methodology focuses in the below points;

- City analysis, climate zones selection,
- Weather data,
- Modeling and simulations,
- Results,

2.1 City analysis, climate zones selection

The city of Dubai has had a rapid in the last 50 years. The main areas develop close to the sea. However, the city is expanding rapidly towards the desert. Mapping such development is quite relevant in defining the most heated areas. There are 3 main highways going in parallel to the coast. It's a linear development with a number of centres on both sides of the highways.
The main LCZ that are also taken into the next step of the study are the urban districts. The rest include the industrial zones, the parks, the water areas, the shrub area and the desert zones. The list of characteristics of the urban zones is mentioned in table 1.
LCZ 1- Compact high-rise, the district analyzed is aligned with Shaikh Zayed Road.
LCZ 2- Compact mid-rise, this are refers to the historic center of Deira.
LCZ 3- Compact low-rise, is located to the area of Karama, close to the downtown area.
LCZ 4- Open high-rise, this zone refer to Jumeirah Lake Towers, where the towers are spread around the central artificial lake living more space betweern each building.
LCZ 5- Open mid-rise, this are refers to the latest developed zones in the Greens.
LCZ 6- Open low-rise, the zone refers to the typical villa compounds like the Irish Village.
LCZ 9- Sparsely Built, this are typically nationals villas with vast space and green between the units. This type of units are spread in different locations all over the city.

Table 1. LCZ Classification for Dubai.

	LCZ 1	LCZ 2	LCZ 3	LCZ 4	LCZ 5	LCZ 6	LCZ 9
District dimensions (m)	574x7960	314x195	293x237	931x612	885x558	362x304	498x517
District area (m2)	188951	61467	69821	570228	494864	107919	258427
No. of buildings	15	29	36	25	23	52	54
Average height (m)	240	24	10	240	24	10	10
Tree %	5	0	6	8	9	10	17

IOP Conf. Series: Materials Science and Engineering **829** (2020) 012013 doi:10.1088/1757-899X/829/1/012013

2.2 *Weather data*

The site measurements are used as boundary conditions for the CFD (Computational Fluid Dynamic) software called ENVI-met. A detailed description is placed in the simulation section. The site data for this study are taken from an open source located in Dubai.
In figure 1 are shown the values of the temperature, Wind speed and wind direction for the week of 20-27 December 2019.

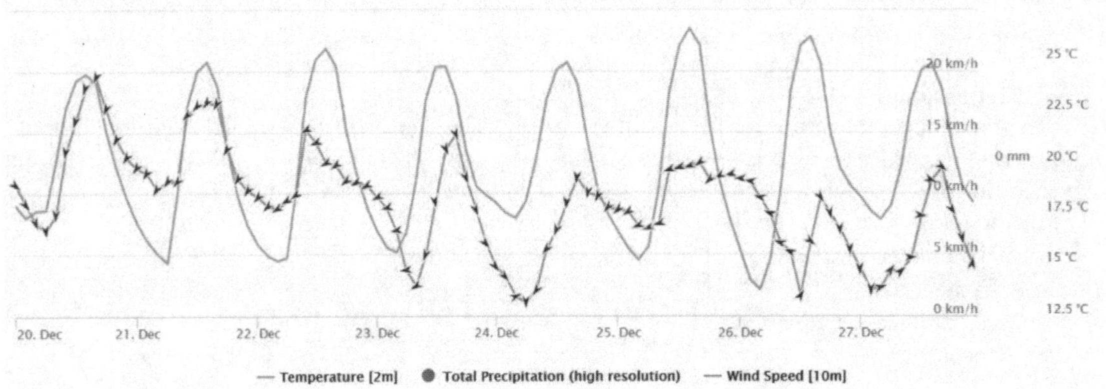

Figure 1. Air temperature, wind speed, wind direction graph, 20-27 December 2019, Dubai.

2.3 *Modelling and simulations.*

Two main softwares are used for the modelling: WUDAPT and ENVI-met. Each tool has a different function. However, the results of ENVI-met should be aligned with the WUDAPT map.

2.3.1 WUDAPT software. Referring to previous studies, the analysis in this analysis is used Level 0 of the LCZ scheme. The LCZ have a correlation with the air temperature. The districts have specific characteristics of the buildings patterns, distance, materials, green zones eth. In order to detect the LCZ first has to be created an initial group of training areas Through a careful study of the heights of the buildings, the land use and land cover areas, the Dubai region can be classified into nine types of buildings and six land cover classes, i.e. a total of 12 LCZ classes, as shown in Table 2. The LCZ classification task is performed using SAGA software on Landsat images after the careful selection of a minimum of three samples of each class. Table 2 shows the outcome of the LCZ for the region of the Dubai City. We make use of Landsat satellite data for LCZ classification. This group is created by an expert that knows the city, such as an architect, urban planner etc., [3][4].

Table 2. Training LCZ of Dubai.

LCZ 1- Compact high-rise	LCZ 2- Compact mid-rise	LCZ 3- Compact low-rise
	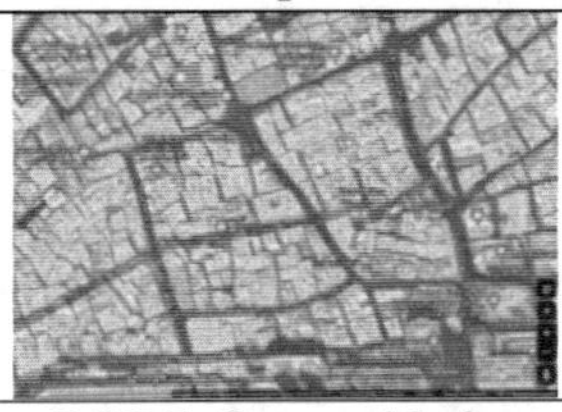	
LCZ 4- Open high-rise	**LCZ 5- Open mid-rise**	**LCZ 6- Open low-rise**

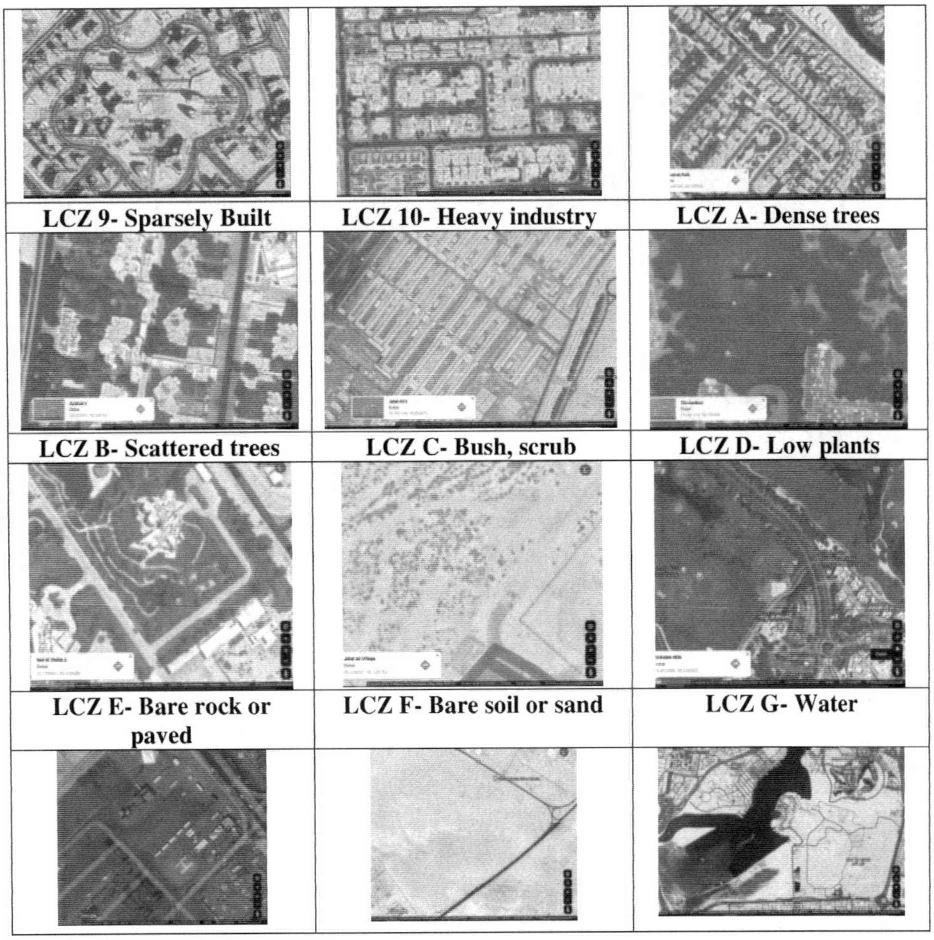

2.3.2 ENVI-met simulations. The 7 selected districts were built with version 4.0. The 3D shown in the table has a telescoping factor greater than 20 m. This makes the 3D images non-realistic. The walls and other surface materials were modified from the original ENVI-met template. This change was made in order to save time in the simulation. In this version of the software, the simulation runs for 24 h.

The trees considered in the simulation are close to the local ones, with small leafs and low irrigation. Meanwhile the soil underneath is irrigating soil so the CFD simulation can go thru the evapotranspiration process. The pavement and asphalt have also been modified in order to have better heat fluxes. With all this modification the model is closer to the real conditions. [10-12].

3. Results

The results are shown in two sections: the WUDAPT results and the ENVI-met results.

3.1. WUDAPT results

Through a careful study of the heights of the buildings, the land use and land cover areas, the Dubai region can be classified into nine types of buildings and six land cover classes, i.e. a total of 12 LCZ classes, as shown in figure 1. The LCZ classification task is performed using SAGA software on Landsat images after the careful selection of a minimum of three samples of each class. Figure 2 shows the outcome of the LCZ for the region of the Dubai City. We make use of Landsat satellite data for LCZ classification.

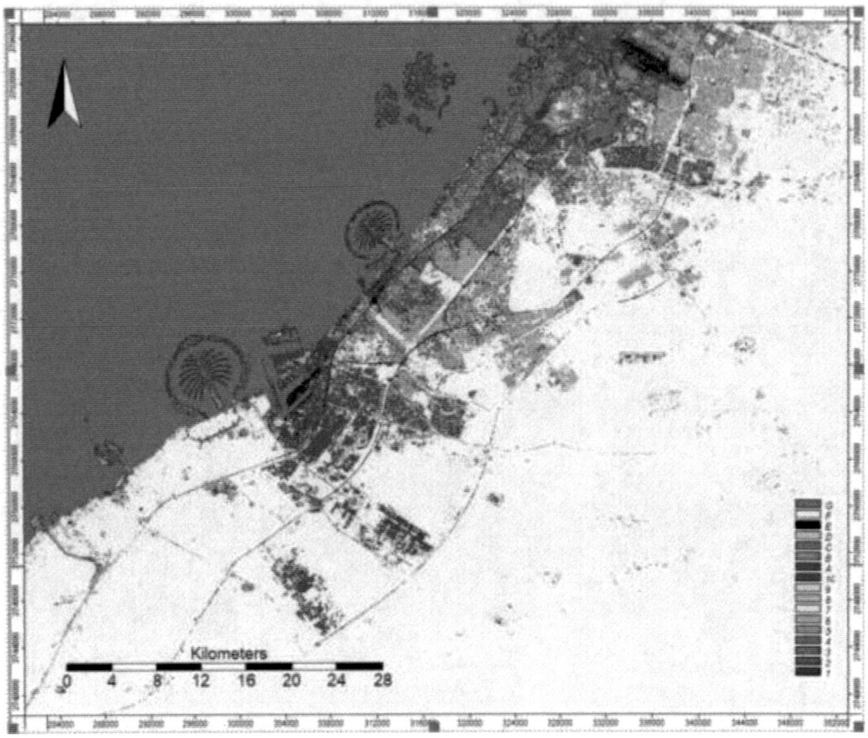

Figure 2. LCZ classification results for the Dubai region.

3.2. ENVI-met results

The ENVI-met results show the different air temperatures at a height of 5 m. The results for a 24 h period were compared with the results from LST. Since a discrepancy is obvious between the simulation results and the measurements of the sensors, we use a fitting model. The model developed to fit the ENVI-met. The simulated Error RMSE =2.73, corrected Error RMSE =1.32

4. Conclusion

This study shows that the results of the LCZ are in alignment with the ENVI-met results. Due to this analysis the city of Dubai has a map of the LCZ therefore of the UHI phenomena. The simulations in the CFD tool are done only for one district due to the time limitation. The aim of the future work is to complete the simulation for the rest of the districts. Also this study refers to one specific season due to the limited site data. The aim is to expand the analysis in the four main seasons of the year. The tools used can be improved according to the latest versions released online.

5. References

[1] I. D. Stewart, T. R. Oke, I. D. Stewart, and T. R. Oke, “Local Climate Zones for Urban Temperature Studies,” *Bull. Am. Meteorol. Soc.*, vol. 93, no. 12, pp. 1879–1900, Dec. 2012.

[2] G. Mills *et al.*, “An Introduction to the WUDAPT project.”

[3] B. Bechtel *et al.*, “Mapping Local Climate Zones for a Worldwide Database of the Form and Function of Cities,” *ISPRS Int. J. Geo-Information*, vol. 4, no. 1, pp. 199–219, Feb. 2015.

[4] L. BANDE, “Heat island mitigation strategies, with focus on the urban shading devices. The case study of Abu Dhabi Main Island, United Arab Emirates.”

[5] L. Bande, A. G. Cabrera, A. Afshari, and M. Martin, “Evaluation of Smart Shading Structures in Mitigating Urban Heat Island in a district of a Hot Arid Climate City (Abu Dhabi).”

[6] A. E. Thomas and A. Elizabeth, “Sustainable landscapes for residential neighborhoods in Dubai: An analysis of the relationship between ambient temperature and water requirements of landscape,” 2013.

[7] H. Radhi, “Evaluating the potential impact of global warming on the UAE residential buildings – A contribution to reduce the CO2 emissions,” *Build. Environ.*, vol. 44, no. 12, pp. 2451–2462, Dec. 2009.

[8] “Middle East Sustainable Cities.” .

[9] Ministry of Environment and Water of UAE, “UAE State of Green Economy 2014,” 2014.

[10] M. Roth and V. H. Lim, “Evaluation of canopy-layer air and mean radiant temperature simulations by a microclimate model over a tropical residential neighbourhood,” *Build. Environ.*, vol. 112, pp. 177–189, 2017.

[11] I. Ozkeresteci, K. Crewe, and M. Bruse, “Cartographic Renaissance’ Hosted by The International Cartographic Association (ICA),” 2003.

[12] S. Huttner and M. Bruse, “NUMERICAL MODELING OF THE URBAN CLIMATE – A PREVIEW ON ENVI-MET 4.0.”

Acknowledgments

The authors express their gratitude to the United Arab Emirates University and Khalifa University for their support during this study.

Factor Analysis of Service Quality of Provincial Contractors to Owner Satisfaction in Construction Projects in Thailand

Grit Ngowtanasuwan

Construction Management Program, Faculty of Architecture, Urban Design, and Creative Arts, Mahasarakham University, 44150 Thailand

grit_n@hotmail.com

Abstract. Service quality in construction projects involves the project owner's perception of the process in terms of interactions, activities, and the attainment of acceptable levels of performance from construction activities. The purpose of this study is to present a factor analysis of service quality of provincial construction contractors in north-east Thailand that influences the satisfaction of project owners, by using Exploratory Factor Analysis (EFA). A total of 132 questionnaires, completed and returned by project owners, were analyzed to confirm the factors. The research found the service quality of the provincial contractors comprised of three factors, namely: (1) empathy, (2) responsiveness, and (3) tangibles, respectively. Guidelines for strategy for improving the service quality of provincial contractors are discussed and presented in the results of this research paper.

1. Introduction

The construction business is a service industry in which servicers (construction contractors) must emphasize the service quality of their work in order to satisfy the project owners and successfully compete for work on future projects [1]. The repeating calls of the project owners for future projects are the most important factors for contractors to ensure sustainability of their business in the construction industry. There are many contractors working in provincial areas in Thailand, but these are largely small enterprises using low technology. In order to sustain their business in the industry, contractors often ask: How can we know and understand which factors relate to the service quality of contractors? Which factors have effects on owner satisfaction and their behavioral intention to offer contractors work on future projects? What strategies should be followed in order to increase the service quality of contractors? As such, the main objective of this research is to study and analyze the factors and variables which relate to the service quality and the project owners' satisfaction of provincial contractors in the Thai construction industry. Results will allow provincial contractors to assess the current status and guidelines for the development of their operations in the industry. The scope of this study is based on an opinion survey in terms of construction service quality of project owners who experienced construction projects with contractors. The projects are located in the central area of north-east Thailand, and include Roi-ed, Khon Kaen, Maha Sarakham, and Karasin provinces. Finally, strategies for the development and support of the service quality of Thai provincial contractors are presented.

2. Literature review

The Cambridge dictionary [2] defines a contractor as a person or company that arranges to supply materials or workers for building construction and performs work on a contractual basis. A contractor is responsible for providing all of the materials, labor, equipment, and services necessary for the construction of a project [3]. A general contractor often hires specialized subcontractors to perform all or parts of the construction work. When using subcontractors, the general contractor is responsible for the quality of all work performed by any and all of the hired workers [3]. Krungsri research center [4] reported that there are some 80,000 construction companies registered for commercial activity in Thailand. Of these, only around 55 qualify as large-scale companies, but an examination of their income shows that these hold a 50% market share measured by income, and that most are registered on the SET (Stock Exchange of Thailand). The three largest contractors, Italian Thai, Ch. Karnchang and Sino-Thai, account for over 50% of the combined income of all construction companies traded on the SET. From this information, there are many construction companies or contractors still operating their construction businesses in Thailand, and these can be classified as small-medium enterprises, using low technology, and who need to improve their service quality. Service quality is generally known as a requirement for attributing success to business firms who have to improve service quality in order to survive in a competitive environment and sustain their customer base. Garvin [5] stated that service quality takes the customer's perspective of quality and focuses on all things that influence the customer's zone of awareness of quality. It is a relativistic evaluation in terms of varying from one customer, relative to another. In construction, service quality involves the customer's perception of the process in terms of interactions, activities, and dynamic events. It also involves the customer's perception of how the work takes place on-site [6]. Quality in the construction industry can be defined as the attainment of acceptable levels of performance from construction activities. This performance would be attained when the activity meets or exceeds the requirements of the project owner. The quality of any product or service is achieved when it conforms to the desired specifications. Parasuraman et al. [7, 8] developed a tool for measuring service quality called SERVQUAL in which service quality is defined as the degree of discrepancy between customers' expectations of a service and their perceptions of service performance (see Eq. 1 below).

Service Quality (SQ) = Customers' Perceptions – Customers' Expectations (1)

SERVQUAL is widely used for surveying customer satisfactions in many service industries. It is a tool used for demonstrating usefulness, and can be used comparatively for benchmarking purposes. SERVQUAL comprise five dimensions, as follows: (1) tangible: physical facilities, equipment, and appearance of personnel. (2) Reliability: ability to perform the promised service dependably and accurately. (3) Responsiveness: willing to help customers and provide prompt service. (4) Assurance: knowledge and courtesy of employees, and their ability to inspire trust and confidence. (5) Empathy: caring; individualized attention the firm provides its customer. These five dimensions, with 22 items of measurement, include: 4 items of tangibility, 5 items of reliability, 4 items of responsiveness, 4 items of assurance, and 5 items of empathy [8]. In this research, SERVQUAL would be used as initial factors of service quality in construction projects. Customer satisfaction is a measure of how products and services supplied by a company meet or surpass customer expectations [9]. Customer satisfaction is defined as the number of customers, or percentage of total customers, whose reported experience with a firm, its products, or its services exceeds specified satisfaction goals [9]. In construction, customer satisfaction means project owner satisfaction. The criteria for selecting a contractor is mainly based on price, the contractor's technical capability, financial capability, and previous experiences of the contractor's competence [9]. Sunindijo et al. [10] performed a research by using SERVQUAL to model the service quality of construction projects in Bangkok and surrounding areas. The research findings showed that the service quality comprised of four dimensions, including reliability, responsiveness, assurance and tangibility, respectively. Fischgrund and Omachonu [11] examined the quality gaps in construction projects by expanding on the previous gap analysis studies by Parasuraman, et al. [7]. The research identified 12 construction service quality gaps that are more nuanced than previous studies. One of the most interesting gaps is that between the project owner's expectations and the project owner's perceptions in construction service quality.

3. Research method

3.1. Identification of factors

The five dimensions with 22 items of measurement from SERVQUAL were used as initial factors and items information. In-depth interviews with experts were conducted within certain topic areas. This technique is designed as a group communication process of experts that aims to conduct detailed examinations and discussions of a specific issue. The process can be continuously iterated until a consensus is determined to have been achieved. This technique was used to confirm the factors and the items. All the factors were summarized, and a group of 3 experts in related fields was chosen, based on their experience, who included: a contractor with more than 20 years' experience in the Thai construction industry, a project owner with more than 20 years' experience, and an expert from an academic institution. All experts carefully and individually considered the listed factors for at least three times per expert until consensus was reached. The initial factors and items were improved and modified for construction service quality with suitability in regard to Thai culture, and one item was added to the list of 22. Once the listed factors were confirmed and categorized by the three experts, they were constructed and made ready for analysis. Factors and items are listed in table 1.

3.2. Questionnaire design

A questionnaire survey was designed for data collection to confirm the influential factors and items of the service quality of the provincial contractors by the project owners. The questionnaire comprised two parts, as follows: (1) demographics of the respondents as respondent's position, type of project, type of project delivery and project value (4 questions). This part was measured by frequency (percentage) of the respondents. (2) 23 items of project owner's perceptions (23 questions) and project owner's expectations (23 questions), totaling 46 questions. This part was measured on a 5-level Likert scale from "strongly disagree" to "strongly agree".

Table 1. Service Quality (SQ) questionnaire items.

Factor	Item
Service Quality: SQ	SQ1: Sufficient equipment and appearing ready to work.
	SQ2: Sufficient manpower to complete assigned work.
	SQ3: Proper material storage and good housekeeping.
	SQ4: Workers dress properly, such as site safety wears.
	SQ5: tails of documents.
	SQ6: Keeping promises.
	SQ7: Showing sincerity when solving problems.
	SQ8: Performing work order as required.
	SQ9: Providing service at the time promised.
	SQ10: Always reporting work progress.
	SQ11: Informing the owner exactly when the service will be performed.
	SQ12: Providing prompt service to the project owner.
	SQ13: Willingness to help the project owner include additional work.
	SQ14: Behaviors of the workers instill confidence to the project owner.
	SQ15: Assuring the quality of work is up to standard.
	SQ16: Politeness to the project owner.
	SQ17: Having the knowledge to answer the owners' enquiries
	SQ18: Having the competence to solve problems correctly
	SQ19: Understanding the owner's specific needs
	SQ20: Giving personal attention to the project owner
	SQ21: Providing maintenance services in the work that has been done.
	SQ22: Tidiness after work.
	SQ23: Operating hours are convenient to the project owner.

3.3. Validity and reliability tests

To ensure the items in the questionnaire were appropriate for data collection, both validity and reliability tests were employed. Validity test: the interviews with the three experts regarding the identification of the factors confirmed the results. The experts reviewed and commented on whether the items were accurate representations to measure the service quality. They also suggested some items which were more appropriate in the context of the research. This exercise was useful in providing content validity and to ensure the items were neither ambiguous nor confusing. Reliability test: Cronbach's alpha was used to evaluate the reliability of the questionnaire. A pilot study was conducted using 30 target samples (provincial contractors in target areas) to determine reliability. The 23 items (5-level Likert scale measurements) were tested by computer statistical software. The outputs gave the Cronbach's alpha coefficient of 23 items in the project owner's expectations of 0.932, and the project owner's perceptions of 0.940, respectively. All coefficients were above 0.7, demonstrating that the questionnaire was reliable [12].

3.4. Data collection

Once the questionnaire was designed, a target group of the project owners was selected using a convenience non-probability sampling technique. The project owners must have had experience in hiring contractors (at least one project) for their construction projects. All of the respondents are project owners in four provinces. The projects are covered in the central area of north-east Thailand including Roi-ed, Khon Kaen, Maha Sarakham, and Karasin provinces, and the data collection period took 3 months. Face-to-face interviews were conducted to explain the details of the questionnaire and ensure that the respondents understood the aim of the survey. In total, 200 questionnaires were completed, but 68 were discarded due to incomplete or biased responses. As such, 132 questionnaires were recognized as valid and analyzed.

4. Results

4.1. Descriptive results

Table 2 shows demographics of the respondents, consisting of respondent's position, type of project, type of project delivery and project value.

Table 2. Descriptive results.

Description	Frequency	Percentage
Respondent's position		
- Project owner	92	69.7
- Owner representative	40	30.3
Type of project		
- Single house	61	46.2
- Town house/Commercial building	23	17.4
- Office building	15	11.4
- Factory/Warehouse	16	12.1
- Apartment building	10	7.6
- Others	7	5.3
Type of project delivery		
- Design-bid-build	86	65.2
- Design-build	46	34.8
Project value		
- Less than1 M Baht	19	14.4

- 1-3 M Baht	50	37.9
- 3-6 M Baht	37	28.0
- 6-10 M Baht	19	14.4
- More than10 M Baht	7	5.3

4.2. Exploratory factor analysis

Exploratory Factor Analysis (EFA) is a statistical method for uncovering the underlying structure of a relatively large set of variables [13]. EFA is used to reduce the number of variables to a smaller set of underlying summary variables, called factors. In this research, EFA was implemented to determine the underlying structure of factors relating to the service quality of provincial contractors to owner satisfaction in construction projects in Thailand. The 23 variables of service quality (SQ) are equal to customers' perceptions, minus customers' expectations for each item. EFA was implemented with varimax rotation using computer statistical software, and the output showed: The Kaiser-Meyer-Olkin (KMO) measure of sampling adequacy = 0.943 (KMO>0.7) Tabachnik and Fidel, 2001) [14]. Bartlett's test of sphericity had a significant value = 0.000 (less than 0.05) with approx. chi-square = 1856.565, df = 253. The extraction of EFA was based on initial Eigen values greater than 1 [15], as shown in table 3 on the following page, and the output analysis showed four components as the number of factors in this EFA. The percentage of explained variance was 60.48%. The EFA output showed that the 32 items were classified into three groups, as follows: factor 1: SQ7, SQ18, SQ6, SQ15, SQ9, SQ21, SQ19, SQ17, SQ22, SQ23 and SQ20 (11 items); factor 2: SQ10, SQ12, SQ11, SQ13, SQ14, SQ16 and SQ8 (7 items); factor 3: SQ3, SQ4, SQ5, SQ1 and SQ2 (5 items) respectively. Furthermore, the reliability of the questionnaire (all respondents) in terms of the three factors was assessed on the basis of Cronbach's alpha coefficient. According to Nunnally [12], a Cronbach alpha coefficient of 0.7 or higher is recognized as acceptable. The values of the coefficient were acceptable for all three factors 1, 2 and 3, ranking from 0.932, 0.883, and 0.808, and the value of all items was 0.883. Table 3 shows the EFA results of factor analysis of service quality of provincial contractors to owner satisfaction in construction projects in Thailand.

Table 3. Factor loading and Cronbach's alpha.

Item	Factor			Cronbach's Alpha	
	1	2	3		
SQ7	0.745	0.165	0.237		
SQ18	0.730	0.156	0.267		
SQ6	0.693	0.198	0.335		
SQ15	0.678	0.365	0.255		
SQ9	0.675	0.349	0.206		
SQ21	0.661	0.328	0.277	0.932	
SQ19	0.653	0.298	0.261		
SQ17	0.637	0.344	0.135		
SQ22	0.630	0.381	0.42		
SQ23	0.609	0.320	0.149		0.953
SQ20	0.582	0.469	0		
SQ10	0.232	0.747	0.297		
SQ12	0.249	0.727	0.228		
SQ11	0.229	0.718	0.263		
SQ13	0.364	0.641	0.235	0.883	
SQ14	0.512	0.584	0.184		
SQ16	0.422	0.505	0.143		
SQ8	0.474	0.502	0.244		
SQ3	0.344	0.159	0.717	0.808	

Table 3. Factor loading and Cronbach's alpha.

Item	Factor			Cronbach's Alpha
	1	2	3	
SQ4	-0.052	0.374	0.711	
SQ5	0.312	0.279	0.704	
SQ1	0.413	0.111	0.575	
SQ2	0.43	0.223	0.532	

Kaiser-Meyer-Olkin (KMO) = 0.943 (KMO>0.7), p-value = 0.000 (Sig.)

In this research, three factors were named by considering the majority of the items included in the factors, namely factor 1: empathy, factor 2: responsiveness, and factor 3: tangibles.

5. Conclusions

This research studied a factor analysis of service quality of provincial contractors to owner satisfaction in construction projects in Thailand. A total of 132 sample questionnaires, completed by project owners experienced in hiring contractors, were analyzed to confirm the factors. Respondent's profiles are shown in table 4 on the following page. Factors relating to service quality in construction were identified and confirmed using the statistical technique of EFA. The research found three factors of service quality of the provincial contractors to owner satisfaction, as follows: (1) empathy: showing sincerity in solving problems, having the competence to solve problems correctly; keeping promises, assuring the quality of work is up to standard; providing the service at the time promised; providing maintenance service in the work that has been done; understanding the owner's specific needs; having the knowledge to answer the owner's enquiries; tidiness after work; operating hours are convenient for the owner; and giving personal attention to the project owner. (2) Responsiveness: always reporting the progress of work; providing prompt service to the owner; informing the owner exactly when the service will be performed; willingness to help the project owner including additional work; behavior of the workers to instill confidence in the project owner; politeness to the project owner; and performing work orders as required. (3) Tangibles: proper material storage and good housekeeping; workers dress properly; site safety implementation; details of documents; sufficient equipment; appearing ready to work; and sufficient manpower to complete the assigned work.
To recommend strategies for improving service quality in construction to the provincial contractors, the top three items in terms of factor loading were selected as a guideline to formulate the strategy for each factor. The recommended factors of service quality of provincial contractors to owner satisfaction in construction projects in Thailand are shown in table 4 below.

Table 4. Recommended factors of service quality of the provincial contractors.

Service Quality	Empathy	- Showing sincerity in solving problems. - Having the competence to solve problems correctly. - Keeping promises.
	Responsiveness	- Always reporting work progress. - Providing prompt service to the project owner. - Informing the project owner exactly when the service will be performed.
	Tangibles	- Proper material storage and good housekeeping. - Workers dress properly, such as site safety implementation. - Details of documents.

IOP Conf. Series: Materials Science and Engineering **829** (2020) 012014 doi:10.1088/1757-899X/829/1/012014

The factors and items in table 4 above could be used as strategies for improving service quality of the provincial contractors to owner satisfaction in construction projects in Thailand, and for sustaining their business in the construction industry.

6. References

[1] G. Arslan, and S. Kivrak: Critical factor to company success in the construction industry, submitted to World Academy of Science, Engineering and Technology, 45 (2008), p. 43-46.

[2] Information on https://dictionary.cambridge.org/dictionary/english/contractor

[3] Information on https://en.wikipedia.org/wiki/Contractor

[4] Information on https://www.krungsri.com/bank/getmedia/7fe1e975-13f4-4040-a276-0d40841e504b/IO_Construction_Contractor_190606_TH_EX.aspx

[5] D.D. Garvin: Quality on the line, submitted to Harvard Business Review, Vol. 61, (1983), p. 65-73

[6] F. Perry, Modelling customer perceived service quality in housing construction, submitted to Engineering, Construction and Architectural Management, Vol .15 Iss 5 (2008), p .485-496

[7] A. Parasuraman, V.A. Zeithaml and L.L. Berry: A conceptual model of service quality and its implications for future research, submitted to Journal of Marketing, Vol .49 No .4, (1985), p . 41-50

[8] A. Parasuraman, V.A. Zeithaml and L.L. Berry: Servqual :a multi-item scale for measuring consumer perceptions of quality, submitted to Journal of Retailing, Vol .64, Spring, (1988), p. 13-40

[9] S. Kärnä: Analysing customer satisfaction and quality in construction-the case of public and private customers, submitted to Nordic Journal of Surveying and Real Estate Research, Vol. 2)2004 :(Special Series .Information on https//:journal.fi/njs/article/view/41488

[10] R.Y. Sunindijo, B.H.W. Hadikusumo, and T. Phangchunun: Modelling service quality in the construction industry, submitted to International Journal of Business Performance Management, Vol. 15, No .3, (2014), p. 262-276

[11] J. Fischgrund and V. Omachonu: Quality in Construction: Identifying the Gaps, submitted to International Journal of Construction Engineering and Management. 3(2) (2014), p. 65-73

[12] J.C. Nunnally, in: *Psychometric Theory* 2nd, New York (1978) , McGraw

[13] M. Norris and L. Lecavalier: Evaluating the use of exploratory factor analysis in developmental disability psychological research, submitted to Journal of Autism and Developmental Disorders, 40(1) (2010), p. 8-20

[14] B.G. Tabachnik and L.S. Fidel, in: *Using multivariate statistics*, 4th ed., Needharm Hights, MA (2001), Allyn & Bacon

[15] H.F. Kaiser: The varimax criterion for analytic rotation in factor analysis, submitted to Psychometrika. Vol. 23, Iss. 3, (1985), p. 187-200

Acknowledgement

This research project was financially supported by Mahasarakham University, Thailand.

IOP Conf. Series: Materials Science and Engineering **829** (2020) 012015 doi:10.1088/1757-899X/829/1/012015

Effect of IR Drop on Reinforced Concrete Corrosion Measurements

U.Raghu Babu[1, a], B.Kondraivendhan[2,b]

[1,2] Civil Engineering Department, Sardar Vallabhbhai National Institute of Technology, Surat, Gujarat, India

[a] ammaraghubabu@gmail.com, [b] kondraivendhan78@gmail.com.

Abstract. The ohmic voltage drop (IR) due to the high resistivity of concrete needs to be considered while monitoring the corrosion performance of rebar embedded in concrete. It helps to evaluate the corrosion condition precisely. Concrete mixes were contaminated with chlorides, sulphates, and a combination of chloride plus sulphate solutions as mix water. To assess the IR drop error in corrosion measurements caused by concrete with different salts, linear polarization resistance values were measured with and without IR compensation. The results indicate that the concrete contaminated with 5% of chloride has shown higher error as compared to 2% of sulphates, due to high corrosion current density. However, from the results of chloride and chloride plus sulphate contaminated concrete, it is also confirmed that the IR drop error is not absolute to determine the solution resistance. The higher corrosion in chloride contaminated concrete as compared to chloride plus sulphate contaminated concrete is also observed due to the availability of high chloride concentration. The results of the present study are suggested that the IR drop compensation should be considered when the reinforced structures are exposed to chloride and sulphate environment.

1. Introduction

Reinforcing steel embedded in concrete is passivated by the ferric oxide film in high alkalinity of concrete[1]. This protective layer gets breakdown, when the chloride ion concentration at the rebar reaches the threshold level and/or the cover concrete is carbonated[2–4]. The competition of hydroxyl ions with chloride ions causes the local adsorption of chloride ions on the passive film and leads to the local dissolution of passive film. The depassivated sites may have lower potential as compared to the other passivated (cathodic) sites on the rebar surface and act as anodic sites. The potential difference along the reinforcing steel in concrete leads to the establishment of an electrochemical cell with anodic, cathodic sites and concrete pore solution as an electrolyte[2]. The following half-cell reactions must take place at the anode and cathodic regions in corrosion phenomenon with the same rate:

- First half cell reaction: The oxidation of iron from Fe to Fe^{+2} releases free electrons. The positively charged ferrous ions pass into the pore solution and free electrons reach the cathodic site through the rebar.
- Second half cell reaction: The liberated electrons from the oxidation of iron are consumed by the cathodic sites and combine with the water and oxygen to form hydroxyl ions

The hydroxyl ions formed by the reduction of oxygen at cathodic region travel through the electrolyte (concrete pore solution) and combine with ferrous ions to form ferrous hydroxide ($Fe(OH)_2$). This

$Fe(OH)_2$ gets precipitated on rebar surface and undergo further reactions to form rust based on the pH value of concrete pore solution, availability of water, and oxygen. It is important to note that the chloride ions present in concrete responsible for localized pitting corrosion of reinforcement, but the neutralization of concrete by the carbonation induces the generalized corrosion.

Half-cell potential, concrete resistivity, and corrosion current density are the main parameters to be considered to evaluate the corrosion state of the rebar in concrete[5]. From the above parameters, the corrosion current density (I_{corr}) only gives quantitative information about the corrosion of rebar in concrete. Electrochemical techniques, namely, linear polarization resistance (LPR), potentiodynamic (Tafel), Galvanostatic pulse, Electrochemical impedance spectroscopy, Harmonic analysis, can be used to determine the I_{corr} values through polarization resistance and Stern-Geary equation[6]. Although I_{corr} value varies based on the adopted technique, LPR technique is considered as most preferable due to its low effect on the rebar condition. The obtained I_{corr} from the LPR technique is purely instantaneous. This technique utilizes the linear relationship of potential (E) and current (I) and at around the corrosion potential (E_{corr}). The slope of E- I curve (polarization curve) at E_{corr} is known as linear polarization resistance (Rp = dE/dI). On the rebar embedded in concrete, the Rp values can be determined by polarizing the rebar with small potential (typically ±20 mV) from $E_{corr.}$ The obtained R_p value is related to I_{corr} using Stern-Geary equation as follows[7]:

$$I_{corr} = B/R_p \qquad (1)$$

Stern-Geary constant B comprises anodic and cathodic Tafel slopes. In the case of rebar in concrete, generally B value is considered as 26 mV in active corrosion state or 52 mV in a passive state of rebar. There is an uncertainty in LPR measurements due to the ohmic drop caused by the finite electrical resistance of concrete and significant distance between rebar and the external electrode[8]. During the flow of polarization current from rebar (working electrode) to the counter electrode, through the concrete causes a potential drop (E_u =IR, where I: polarization current, R: uncompensated resistance). This potential drop is generally referred to as an ohmic drop (IR), electrolyte resistance or solution resistance. This voltage drop exists during the electrochemical measurements on rebar embedded in concrete due to the flow of current through finite resistivity of concrete. The corrosion measurements ascertained without considering the ohmic resistance leads to too small I_{corr} values, which are erroneous[9]. The importance of ohmic drop in cathodic protection is also significant. In cathodic protection, corrosion is reduced by polarizing the cathodic sites towards the anodic potential. When the cathodic protection current is flowing through the soil/concrete, the IR drop caused by an applied current within the cathodic region supports in reducing the cathodic protection current[10]. Having in mind the flow of current through concrete, it appears reasonable that electrical conductivity of concrete pore solution, microstructure, moisture content and temperature are playing a role in IR drop. However, the pore solution chemistry depends on pH and ionic species like chlorides, sulphates, and other alkali metal ions.

As the problem due to IR drop is well known, several IR compensation methods have been developed. Nevertheless, the problem is still neglected by many users, especially for the corrosion measurements of reinforced concrete structures in the field. Keeping this in view, an attempt has been made to estimate the IR drop error level in LPR measurements to highlight the importance of IR compensation in corrosion studies on reinforced concrete structures.

In this work, the error level due to the IR drop in LPR measurements is investigated. The IR drop errors of various samples are discussed concerning the admixed salt type like chloride, sulphate.

2. Experiment

2.1 Materials and specimen preparation

Three concrete mixtures were prepared with three different admixed solutions and water-cementitious material (w/cm) ratio 0.48. River sand, which complied with Zone II and natural coarse aggregate with a nominal maximum size of 20 mm, were used in this study. As well as ordinary Portland cement meeting Indian standards IS: 12269 -2013[11] was used. The concrete mixtures were contaminated with 5% of NaCl as chloride source, 2% of $MgSO_4$ as sulphate source, and a combination of both the salts. The salts were added in mix water while preparing the concrete mixtures. Rebar of diameter 10mm was embedded centrally in concrete slab specimens of size 320mm x 320mm x 52 mm. Rebar was cleaned to remove all the rust on the surface of the rebar before placing it into the concrete. The two ends of the rebar were protected with a neoprene tube such that the middle portion of length 250mm is exposed to monitor the corrosion measurements. After demoulding, the reinforced slab specimens were exposed to curing for 28 days, after that exposed to normal ambient laboratory conditions for two months.

2.2. Corrosion monitoring technique

Corrosion monitoring instrument ACM (Advanced Corrosion Monitoring, U.K) Gill AC guard ring serial no.1824-sequencer was used to determine the corrosion current density of the rebar in slab specimen. Before testing, the slab specimens were immersed in the water for 10 hours to avoid the noise caused by too much dryness. The LPR technique was used to monitor the I_{corr} values on slab specimens using the guard ring. The other details about the guard ring assembly were discussed elsewhere[3]. Rebar in slab specimen was polarized to ±20 mV from the corrosion potential with the scan rate of 0.1 mV/second. The LPR technique was performed with and without the IR compensation technique.

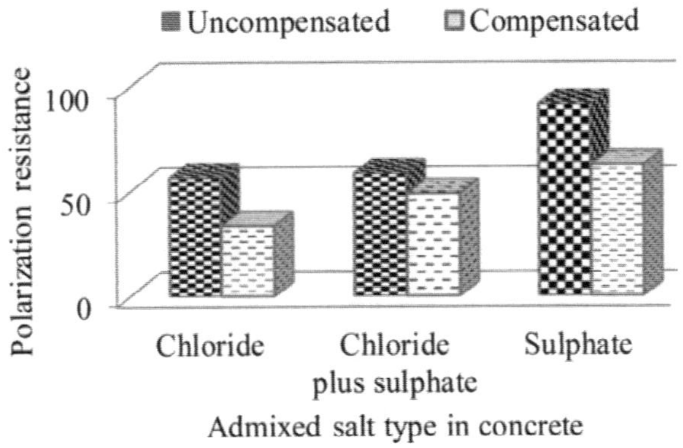

Figure 1. Polarization resistance with and without IR drop compensation LPR technique.

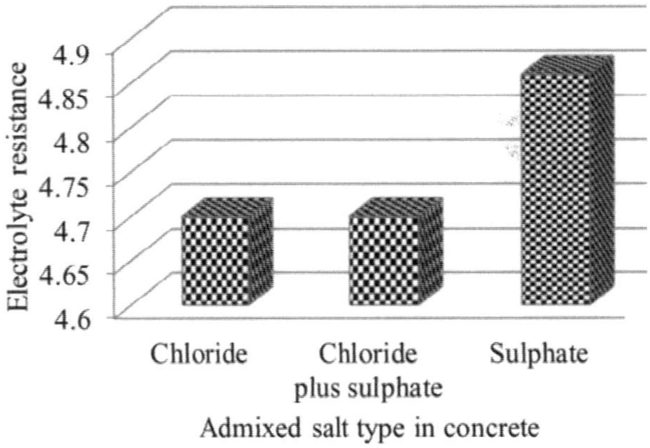

Figure 2. Electrolyte resistance of concrete mixes contaminated with different salts.

3. Results and discussion

In the process of polarization current flow from counter electrode to rebar through the concrete, there is a resistance in current flow, which may reduces the actual polarization voltage of the rebar in concrete. The measured polarization resistance ($R_{P,M}$) calculated by uncompensated LPR technique is equal to the sum of real polarization resistance (R_P) and IR drop resistance (R)[12]. Due to the presence of IR dro, $R_{P,M}$ value leads to very high and therefore a measured corrosion current density ($I_{corr,M}$) will be too low:

$$I_{corr,M} = B/R_{p,M} = B/(R_P + R) \quad (2)$$

The relative error in the measurement of corrosion current density value can be determined by the ratio of IR drop resistance and polarization resistance. This can be indicated as follows:

$$\Delta = (I_{corr} - I_{corr,M})/ I_{corr,M} = R/R_p \quad (3)$$

To demonstrate the error level caused by the uncompensated polarization, the results obtained by the laboratory experiments have been discussed in this section. Figure 1 depicts the differences between R_p and $R_{p,M}$ values, which are measured by compensated and uncompensated LPR techniques, respectively. figure 2 and figure 3 shows the solution resistance and relative errors of different concrete mixes made with chlorides, chlorides plus sulphates and sulphate solutions. From figure 1, it can be observed that the $R_{p,M}$ values are higher than the R_p values. It is due to the well-known fact that the uncompensated $R_{p,M}$

accounts both R_p and solution resistance. As the minimum R_p value represents the higher corrosion, from figure 1, it can be observed that the concrete admixed with chloride shows higher corrosion as compared to the concrete admixed with chlorides plus sulphate solution. While observing the uncompensated polarization value (i.e., $R_{p,M}$ values), the difference between chloride samples and chlorides plus sulphate samples is not significant, but it is more significant in compensated resistance. While observing figure 2, the resistance of cover concrete or electrolyte resistance is approximately the same between the chloride and chloride plus sulphate contaminated concrete samples, although the compensated polarization resistance (R_p) is varied. It can be an important observation since it explains that the error is not an absolute value of solution resistance. Oldham et al. [13]reported that the error could be substantial even in high conductive electrolyte solution if the polarization resistance value is low. The high corrosion of rebar in chloride contaminated concrete is attributed to the availability of high chloride concentration as compared to the concrete contaminated with chlorides plus sulphate solution[14]. The polarization resistance and electrolyte resistance values are higher in concrete, admixed with magnesium sulphate solution alone. This can be attributed to the low electrical conductivity of sulphates present in concrete[15].

Figure 3 shows the corrosion current density values measured by both compensated and uncompensated LPR technique. Due to the reciprocal relation between Rp and I_{corr} values, I_{corr} values are increased with decrease of R_p. It is observed that the concrete contaminated with chlorides has shown the higher corrosion and the concrete contaminated with sulphates has shown lower corrosion. It confirms that the participation of sulphates alone in corrosion mechanism is negligible.

It is also observed that $I_{corr,M}$ of rebar in concrete contaminated with chlorides and sulphates is not similar with I_{corr} values, which are obtained by compensated LPR technique. It is evident that the effect of IR drop is most considerable in corrosion monitoring in reinforced concrete structures. The error due to IR drop can be illustrated clearly with the difference between corrosion current density values measured by both compensated and uncompensated technique, as shown in figure 4.

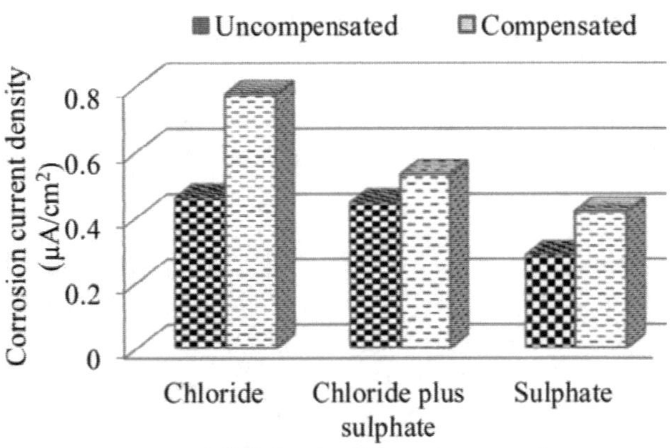

Figure 3. Corrosion current density values with and without IR drop compensation LPR technique.

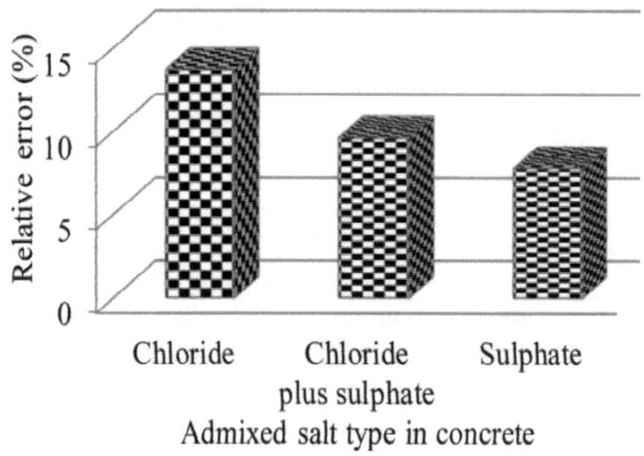

Figure 4. Relative error in corrosion measurements due to IR drop in concrete contaminated with different salts.

The relative error in chloride, chloride plus sulphate and sulphate admixed concrete mixes is 14%,9.6% and 7.8%, respectively. High relative error was observed at high corrosion values, that is the rebar in chloride contaminated concrete mix. However, the less relative error was observed in sulphate contaminated concrete specimens, in which the corrosion is less. From this observation, it is clear that the IR compensation needs to be considered when the corrosion of reinforcement is high.

4. Conclusion

1. Rp and $R_{p,M}$ values measured on chloride and chloride plus sulphate contaminated concrete are confirmed that the IR drop error is not an absolute value to explain the solution resistance of the sample.
2. The IR drop error is more for chloride contaminated concrete as compared to other concrete contaminated with combined chloride plus sulphate and pure sulphate solutions, due to the high corrosion rate in a chloride environment.
3. As compared to chloride contaminated concrete samples, the sulphate contaminated samples show less IR drop error due to the low electrical conductivity of sulphates.
4. The high corrosion in chloride contaminated concrete as compared to the chloride plus sulphate contaminated concrete is due to the availability of high chloride concentration.

5. References

[1] Yang L, Xu Y, Zhu Y, Liu L, Wang X and Huang Y 2016 Evaluation of Interaction Effect of Sulfate and Chloride Ions on Reinforcements in Simulated Marine Environment Using Electrochemical Methods *Int. J. Electrochem. Sci.* **11** 6958

[2] Ismail MAE 2003 *Corrosion of reinforcement - An Application of concrete technology from theory to practice*

[3] Raghu babu U and Kondraivendhan B 2019 Impact of sulphate on chloride-induced corrosion of steel in concrete *Indian Concrte J.* 8–17

[4] Raghu Babu U and Kondraivendhan B 2020 Application of Statistics to the Analysis of Corrosion Data for Rebar in Metakaolin Concrete Learning and Analytics in Intelligent Systems vol **2**, ed S C Satapathy, K S Raju, and G A Tsihrintzis (Cham: Springer International Publishing) pp 162–9

[5] Palaniswamy N and Vedalakshmi R 2010 Analysis of the electrochemical phenomenon at the rebar–concrete interface using the electrochemical impedance spectroscopic technique *Mag. Concr. Res.* **62** 177–89

[6] Vedalakshmi R and Thangavel K 2011 Reliability of Electrochemical Techniques to Predict the Corrosion Rate of Steel in Concrete Structures *Arab. J. Sci. Eng.* **36** 769–83

[7] Stern M 1957 Electrochemical Polarization *J. Electrochem. Soc.* **104** 559

[8] Walter G W and Centre T 1977 Problems arising in the determination of accurate corrosion rates from polarization resistance measurements **17** 983–93

[9] Oelßner W, Berthold F and Guth U 2006 The iR drop – well-known but often underestimated in electrochemical polarization measurements and corrosion testing 455–66

[10] Angst U M and Büchler M 2018 Impact of IR drops on the -850 mvcse cathodic protection criterion for coated steel pipes in soil *J. Pipeline Syst. Eng. Pract.* **9** 1–11

[11] Bureau of Indian Standards 2013 Ordinary Portland Cement, 53 Grade-Specification (First Revision) *IS 122692013*

[12] Xu J, Cao Y, Jiang L, and Feng W 2016 Error Estimation of Linear Polarization Resistance Measurement Caused by IR Drop for Reinforcement Corrosion in Concrete **866** 14–9

[13] Oldham K B and Mansfeld F 1973 Corrosion rates from polarization curves: A new method *Corros. Sci.* **13** 813–9

[14] Shaheen F and Pradhan B 2015 Effect of chloride and conjoint chloride – sulfate ions on corrosion of reinforcing steel in electrolytic concrete powder solution (ECPS) *Constr. Build. Mater.* **101** 99–112

[15] Sipilä K, Bojinov M, and Stanislowski M 2015 Effect of chloride and sulfate additions on corrosion of low alloy steel in high-temperature water *Electrochim. Acta* **173** 757–70

Evaluation of Mechanical Hardened Properties of Mortar Using Carbon-Free Fly Ash and Normal Fly Ash

Ghawsaddin Nazari[1], Sato Takayuki and Shigeyuki Date[3]

[1]4-1-1 Kitakananme, Hiratsuka, Kanagawa, Japan
[2]1-1-10 Itsutsubashi Aobaku, Sendai, Miyagi, Japan
[3]4-1-1 Kitakananme, Hiratsuka, Kanagawa, Japan

[1] Ghm255.b@gmail.com

Abstract. Usage of mineral admixture and chemical admixture in concrete or mortar is a usual solution to reach full compaction, particularly where reinforcement blockage and lack of skilled labour happen. In this paper effect of mineral admixtures (Carbon-free fly ash, hereafter CfFA, and nFA) on hardened properties of mortar such as compressive strength, Pore size distribution and, and micro structures have been investigated. As a result, it has been confirmed that mortar containing 15% CfFA slightly increased the compressive strength under water curing condition compared to those containing 15% normal fly ash. In addition, the pore-size distribution of the mortar has been affected by variation of materials and curing condition. Mortar containing 30% CfFA had a smaller threshold diameter in comparison to that of 30% FA; however, there is no significance differences in mortar containing 15% CfFA and FA. Furthermore, CfFA-based mortar increased the amount of entrained air compared to the mortar with normal fly ash.

1. Introduction

Concrete is the most widely used construction material in the modern world. Its huge popularity is due to the result of many advantages: such as low cost, general availability, and wide applicability. However, it has a notable climate impact due to the large volumes and the energy intense production of cement clinker which has a GHG emission between 5 and 8 % of the world's total. [1, 2]. On the other hand, usage of mineral admixture as partial replacement of cement such as silica fume, fly ash (FA), bottom ash (BA), ground granulated blast furnace slag (GGBFS) and others have increased. [3-7]. These materials not only lead to reduction CO_2, development of fresh and mechanical harden properties of concrete, but also utilize the industrial by-product which needs more landfill and waste disposal.

Among them, fly ash which is a by-product of the combustion of pulverized coal in electric power generating plants is the most widely used SCM in concrete that is used in more than 50 percent of ready-mixed concrete. Appropriate use of FA prevents expansion due to alkali – silica reaction, reduces heat generation and gives better durability in concrete. [8,9]. However, the presence of carbon influences fresh properties of concrete. The carbon content affects the air entraining agents and diminishes the entrained air for a given amount of air-entraining agent. [10,11]. In addition, High volume fly ash concrete hydrates more slowly than ordinary concrete. [12,13]. This factor, which increases with increasing fly ash replacement dosage, causes problems in concrete construction where rapid stripping and demoulding are essential.

Table 1. Materials used.

Material		Description	Density (gr/cm^3)
Cement	OPC	Ordinary Portland Cement	3.16
	nFA	Normal fly ash (grade 2 fly ash)	2.50
	CfFA	Carbon free fly ash	2.20
Fine aggregate	S	River sand From Yamakita Kanagawa	2.69
Superplasticizer(sand)	SP	Polycarboxcylic –acid based	-
Air Entraining agent	AEA	Polycarboxcylic –acid based	-
Powder	P	OPC+ (CfFA or nFA)	-
Water	W		1.0

Table 2. Mix proportions.

CfFA & FA replacement (%)		W/P (%)	S/P	SP/P (%)
CfFA	15	30	2.0	0.8~1.0
	30			
nFA	15			1.2~1.3
	30			

Table 3. Chemical composition of CfFA and nFA.

CfFA	nFA	Chemical Composition
64.6	65.5	Si (SiO2)
21.7	21.5	Al (Al2O3)
8.3	7.9	Fe (FeO)
3.6	3.4	Ti (TiO2)

2. Experiment summary

2.1. Materials used and mix proportions of mortar

The materials used in this study are shown in table 1. Referring to table 2, ordinary Portland cement (OPC) was partially replaced with CfFA and normal FA, 15% and 30% respectively. Two type of chemical admixture are used, superplasticizer, PCE based, and air entraining agent. The water-cement ratio was set to 30%, and the sand-binder ratio was set to 2.0. The sand was on saturated surface dry condition from River. Using SP, the flow was adjusted to be within the range of 140±20mm and 200±mm by 15 tamping, and the air content was adjusted to 7%±0.5. The chemical composition of carbon-free fly ash and normal fly ash are also given in table 3.

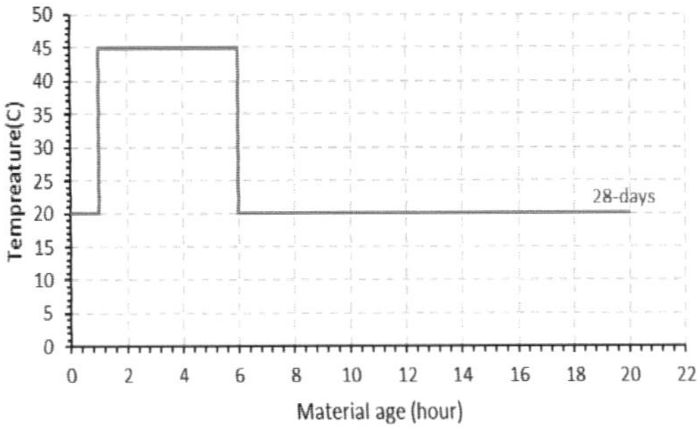

Figure 1. Steam curing pattern.

2.2. Mortar mixing method

The mixing procedure was according to "JIS R 5201" physical testing methods of cement". The OPC was replaced with CfFA and FA. Initially, cement, sand and carbon free fly ash were mixed on a slow speed of mixer for 30 second, subsequently, water, superplasticizer and air entraining agent were added and mixed or 60 second on slow speed as well, and finally again mixed for 30 second on fast speed. The same procedure was applied for fly ash.

2.3. Curing method
In this study, two different type of curing have been conducted: water curing and steam curing. In the water curing the specimens were kept under water for 28 days and 20°C. In steam curing, the specimens were kept in the ambient temperature for one hour at 20 °C and then steamed at 45 °C for 6 hours. After the steam curing has been done, again the specimen kept in the ambient temperature for 28 days. The detail of steam curing pattern is shown in figure 1.

2.4. Test items and test method.
2.4.1 Compressive strength. The compressive strength test was conducted in accordance with (JIS A 1108) The specimens (cylinder with 100mm height and 50mm diameter) were demolded 24 hours after casting. Consequently, the specimens with steam-cured were kept in the ambient temperature for 28 days while other specimens were kept under water for this age. After that, they were crashed.
2.4.2 Pore-size distribution using mercury intrusion prosimetry. MIP tests were performed on mortar samples after each curing for 28 days. The equipment used is an Autopore III 9420 from Micromeritics that can generate a maximum pressure of 414 MPa (60000 psi) and can evaluate a theoretical pore diameter of 0.003μm. With this machine, the MIP test is performed in two steps: the low-pressure step first evacuates gases, fills the sample holder with mercury and performs porosimetry from about 7 to 345 kPa; the high-pressure step reaches pressures between 345 kPa and 414 MPa. Low and high-pressure steps were always performed within a period of 8 h. The contact angle and surface tension assumed for all tests were respectively 130° and 485 dyn/cm.
2.4.3 X-Ray CT scanner and CT images. The μ-Focused X-Ray CT scanner is used to observe the micro structure in the specimen. An illustration of inner view and its specifications is shown in figure 2. A flat panel detector (FDP) is applied to perform a three-dimensional scan with an X-Ray Cone beam. Once the specimen was set up on the specimen table, a cone-shaped beam is emitted from the X-Ray tube and the specimen is scanned. During the scanning, only specimen table is rotated so that the X-Ray beam can penetrate from 360 degree. The scan condition in this study: voltage is 195KV, current is 160μA slice pitch 32.13μm and slice thickness is 32.13μm. Therefore, the voxel dimension is (32.13 x 32.13 x32.13μm.

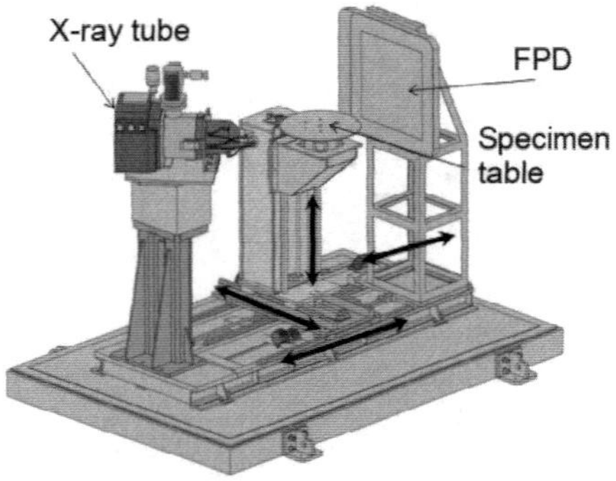

Figure 2. Schematic of the μ-focus X-Ray system.

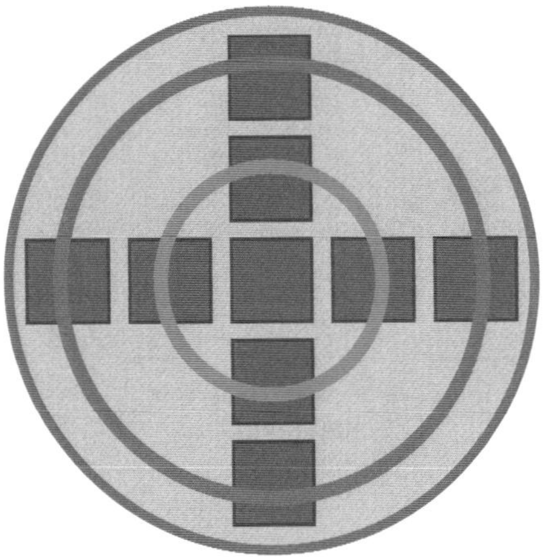

Figure 3. Location of the cell in the specimen.

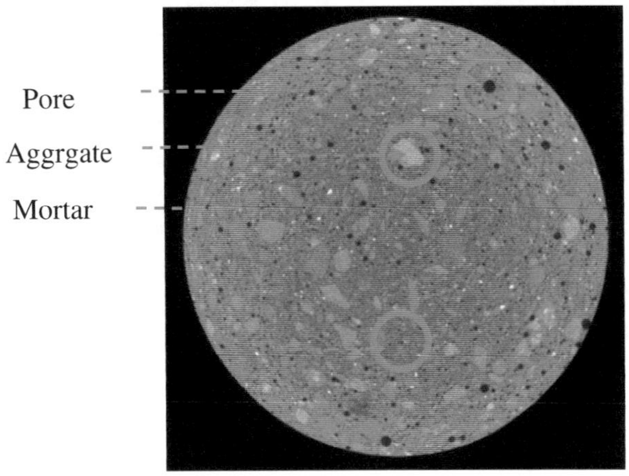

Figure 4. X-Ray CT image.

As it is shown in figure 3, the specimen has been divided into three parts: center, inside and outside, and the volume of the selected cell was 8x8x8mm. The amount of entrained air has been calculated from center, inside and outside separately.

Figure 4 shows a two-dimensional X-Ray CT image of a specimen with diameter of 50mm. The concrete specimens consist of fine-aggregate, mortar and pore. Those are distinguished in the CT images by differences in the light and shade. The high-density aggregate shows light gray, pores with low density are black, and mortar with medium density show as dark gray.

3. Results and discussion

3.1. Comparison of compressive strength of mortar containing different amount of CfFA and FA.

Figure 5 and figure 6 represent the compressive strength of mortar in which cement has been supplemented by various amount of Carbon-free fly ash and normal fly ash. Considering 15% replacement, it was confirmed that CfFA increased the compressive strength of water-cured mortar after 28 days compared to FA; however, there is no significant change after 7 days. In contrast, under the steam-curing condition, FA 15% showed better compatibility after 28dyas, and had greater compressive strength than CfFA 15%.

Referring to figure 6, again CfFA 30% increased the compressive strength after 28 days under water-curing condition compared to FA 30%; however, under steam-curing condition no change can be observed. It is mentionable that Fa 30% had higher compressive strength after 7days than CfFA 30%.

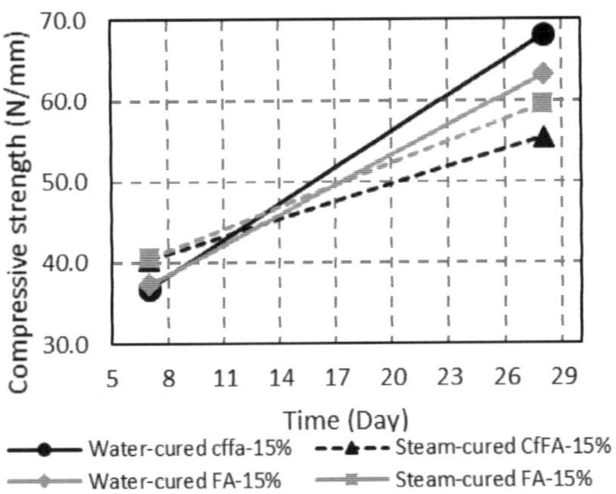

Figure 5. Compressive strength of mortar containing 15% CfFA and FA.

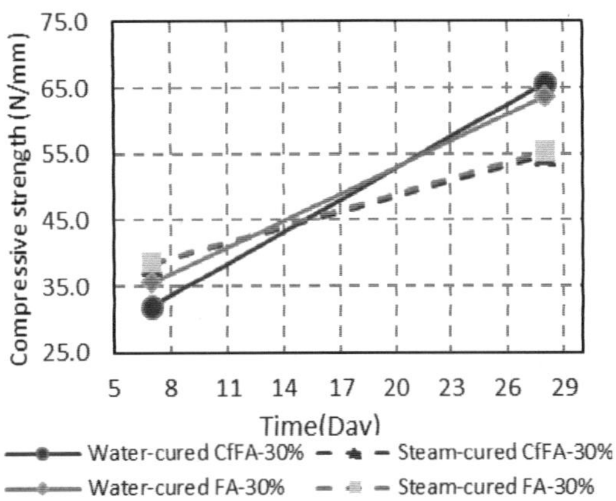

Figure 6. Compressive strength of mortar containing 30% CfFA and FA.

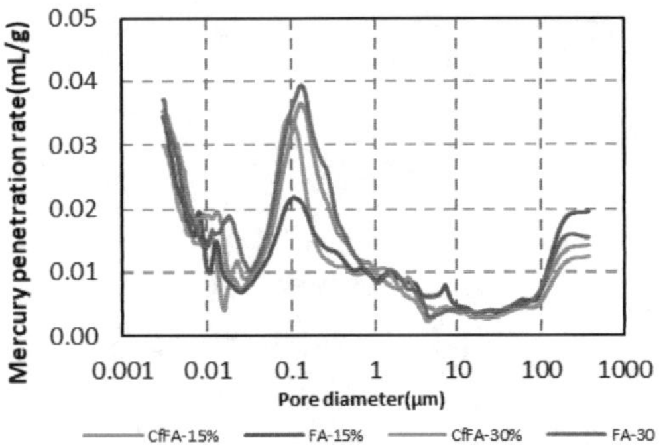

Figure 7: Mercury penetration rate vs pore diameter.

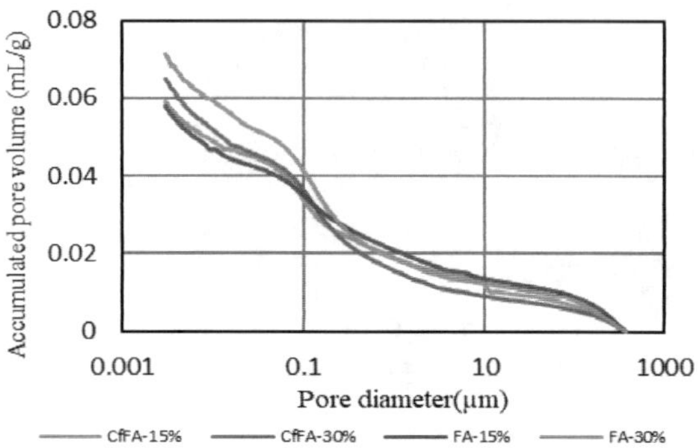

Figure 8: accumulated pore volume vs pore diameter.

3.2. Evaluation of pore-size distribution of mortar containing different amount of CfFA and FA using MIP

In this section, the pore structure of mortar with different mix proportion after 28 days of curing was discussed. Figure 7 explains the cumulative intrusion, reflecting the connected pore volume of pore size larger than specified diameter and figure 8 represents the incremental intrusion at each pore diameter and the reflected pore volume at that as well.

As it is seen, the total mercury intrusion and the threshold diameter value are different in each type of mixing. The lager threshold diameter (TD) can be observed in 30% fly ash content mortar, by 0.15μm, as the total mercury intrusion reaches to 0.07 ml/g. in contrast, the TD value has been shifted to 0.09μm in 30% CfFA content mortar; however, total mercury intrusion was slightly changed. In addition, there is no significant change in the amount of total mercury intruded into 15% CfFA and 15% FA; however, the TD is different.

3.3. Evaluation of pore-size distribution of mortar containing different amount of CfFA and FA using X-Ray Ct test

Figure 9 and figure 10 represent both the amount of entrapped air and entrained air in the mortar partially replaced by 15% and 30% CfFA and normal fly ash. Considering the 15% replacement, it has been confirmed that the CfFA not only increased the amount entrained air in the mortar, but also it has the tendency to increase toward the center. In contrast, normal fly ash had a lower value of entrained air and also the value tend to decline toward the center.

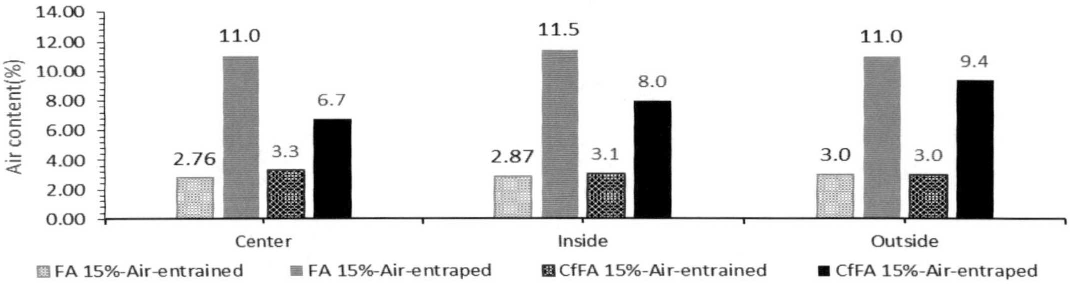

Figure 9. Percentage of entrapped and entrained air in mortar containing 15% CfFA and 15% FA.

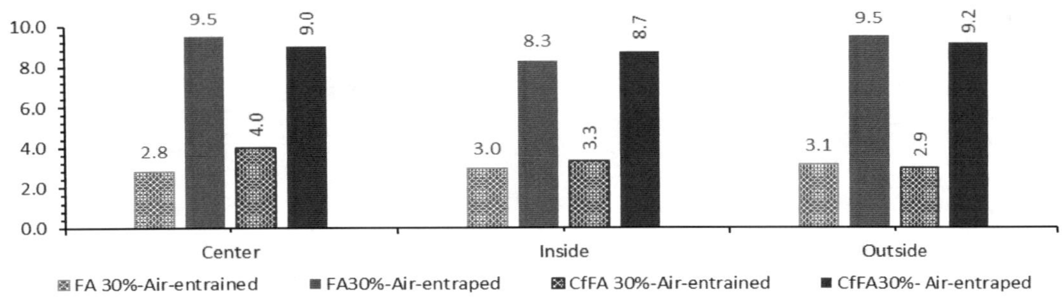

Figure 10. Percentage of entrapped and entrained air in mortar containing 30% CfFA and 30% FA.

This comparison could be observed in 30% replacement. Referring to the center, the amount of entrained air in CfFA is almost twice as much as in normal fly ash. This difference has slightly been changed in the location of inside of the mortar. in CfFA, entrained air tended to decline toward the edges; however, by the percentage of 3.3%, still CfFA had a greater value of entrained air than normal fly ash. in contrast, in the location of outside, the amount of entrained air is higher in normal fly ash than CfFA.

4. Conclusions

The following conclusions has been derived from the experimental study of effect of carbon-free fly ash and normal fly ash on harden properties of mortar:

1-CfFA slightly increased the compressive strength of mortar after 28 days of water curing compared to FA; however, there is no significance change after 7 days.
2-CfFA-15% had smaller threshold diameter (TD) compared to FA-15%.
3-The accumulated pore volume in CfFA-30% was lower than FA-30%.
4-The percentage of entrained air is higher in CfFA than FA, both 15% and 30%.

5. References

[1] International Energy Agency, Cement technology roadmap 2009 - Carbon emissions reductions up to 2050, http://www.iea.org/publications/freepublications/publication/Cement.pdf.

[2] Olivier, J.G.J., Janssens-Maenhout, G., Muntean, M., and Peters, J.A.H.W., , Trends in global CO_2 emissions; 2016 Report, The Hague: PBL Netherlands Environmental Assessment Agency; Ispra: European Commission, Joint Research Centre, 2016.

[3] H.A. Toutanji, Z. Bayasi, Effect of curing procedures on properties of silica fume concrete, Cem. Concr. Res. 29 (1999) 497 – 501.

[4] N. Bouzoubaa, M. Lachemi, Self-compacting concrete incorporating high volumes of class F fly ash preliminary results, Cem. Concr. Res. 31 (2001) 413 – 420

[5] Y.M. Zhang, W. Sun, H.D. Yan, Hydration of high-volume fly ash cement pastes, Cem. Concr. Compos. 22 (2000) 445 – 452.

[6] M. Cheriaf, J.C. Rocha, J. Pera, Pozzolanic properties of pulverized combustion bottom ash, Cem. Concr. Res. 29 (1999) 1387 – 1391.

[7] I. Kula, A. Olgun, Y. Erdogan, V. Sevinc, Effects of colemanite waste, coal bottom ash and fly ash on the properties of cement, Cem. Concr. Res. 31 (2001) 491 – 494.

[8] C.S. Poon, L. Lam, Y.L. Wong, A study on high strength concrete prepared with large volumes of low calcium fly ash, Cem. Concr. Res. 30 (2000) 447 – 455

[9] M.H. Shehata, M.D.A. Thomas, The effect of fly ash composition on the expansion of concrete due to alkali – silica reaction, Cem. Concr. Res. 30 (2000) 1063 – 1072

[10] U.S.dept of Transportation Federal Highway Administration: www.fhwa.dot.gov

[11] ACI material Bulletin E3-01: Cementitious Materials for Concrete.

[12] P. Seabrook, K. Campbell, Levelton Engineering Ltd: Sustainability in Construction: Using Fly Ash as a Cement Replacement. Journal of the Association of Professional engineers and Geoscientist of BC, September 2000
[13] A. Bilodeau, V. M. Malhotra, and P. T. Seabrook, Materials Technology Laboratory: Use of High-Volume Fly Ash Concrete at the Liu Centre. 2001. (Class F fly ash)

Acknowledgement
This research was supported by the Nippon paper corporation and civil and architectural department of the Tokai University.

Cyclic Tests of Joints of Glued Wooden Structures

L B Aruova[1*], Z N Ospanova[2], B B Aruov[3], N T Alibekova [4], Zh A Shashpan[5], A T Kyrgizbaev[6]

[1]Doctor of Technical Sciences, Professor, Eurasian National University named after L.N. Gumilev, Kazakhstan Republic
[2]Candidate of Technical Sciences, Associate professor, Eurasian National University named after L.N. Gumilev, Kazakhstan Republic
[3]Candidate of Technical Sciences, Associate professor, S.Seifullin Kazakh AgroTechnical University, Kazakhstan Republic
[4]Doctor PhD, Associate professor, Eurasian National University named after L.N. Gumilev, Kazakhstan Republic
[5]Doctor of Technical Sciences, Professor, Eurasian National University named after L.N. Gumilev, Kazakhstan Republic
[6]Candidate of Technical Sciences, Associate professor, Eurasian National University named after L.N. Gumilev, Kazakhstan Republic

[*] ecoeducation@mail.ru

Abstract. The use of large-span structures made of glued wood in countries with large timber reserves in Finland, Sweden, Norway, Russia, the USA and Canada has shown that this can be quite interesting and expressive from the architectural position of the building. Sports facilities, sports and sports facilities, sports facilities, sports and cultural facilities. The practice of building bridges made of glued wood in these countries shows that they have been in operation for decades without additional operating costs. Glued wood in modern conditions, when productive flame retardants are used that protect it from fire, antiseptics from possible decay expand the scope of such structures. One of the important obstacles to the limiting applications of large-span glued wooden structures is the complexity of the solution nodes. Bearing structures experience difficulties when exposed to static and cyclic loads. Cyclic loads can be caused by wind loads, production equipment, traffic, and seismic processes. Reinforcement in wood, reinforcement and reinforcement under load. Wood and metal are resistant to cyclic loads. Studies will be conducted at the border of the elements to be glued. This article presents the results of tests for the effects of static and cyclic loads on reinforcement from a 14 mm reinforcing bar of class A5 on FRF-50T along natural-sized glued wood fibers. Certain coefficients of the endurance of the compound when exposed to cyclic loads with two load asymmetry coefficients p = 0.5 and p = 0.2.

1. Introduction

Laminated wood is increasingly being used as the main load-bearing structure for public buildings, including for entertainment purposes. The use of large-span structures in Europe and Canada from glued wood has shown in practice that it can be quite interesting and expressive from the architectural position of the building. Glued wood in modern conditions, when productive flame retardants are used that protect it from fire, antiseptics from possible decay expand the field of application of such structures [1,2].

IOP Conf. Series: Materials Science and Engineering **829** (2020) 012017 doi:10.1088/1757-899X/829/1/012017

One of the important obstacles limiting the use of designs - the complexity of the solution nodes. Since the 1980s, research has been actively conducted in Europe on the development of units of wooden structures on glued and screwed steel rods of long-span structures [3]. Climate change in recent decades causes a significant variety of disasters in nature, it is the temperature changes that cause an increase in wind load in the form of tornadoes, storms, hurricanes. For example, every year the number of hurricanes along the southeastern coast of the United States increases and leads to the destruction of the first coating of private homes. Studies of the work have shown that the destruction of the roof from the effects of the hurricane occurred on the walls of panels made mainly on nails [4]. Studies of structures and their components on the effects of static loads are not sufficient, since a number of structures are experiencing dynamic loads. For example, for bridge structures made of laminated wood, the impact of dynamic loads resulting from uneven movement of load-lifting vehicles along one of the tracks is especially important [5]. Three types of joints are used in wooden structures: connections on cuttings-frontal, in a half-tree and others; mechanical connecting parts using bolts, nails, bands and brackets, metal fasteners - lining, gussets and other; adhesive joints, which allowed in the production of wood products to produce structures of considerable size, any configuration and any size.
The third type of compounds on adhesives allowed to revive wood as a constructive material in the field of construction of public buildings. The only obstacle is the complexity of solving nodes. In the design and application of large-span structures due to the impossibility of transportation and installation, large structures must be divided into elements that in the construction conditions must be connected. These compounds are subject to stringent requirements for strength, compliance, fire resistance and durability.

2. Experiment objectives

In this work, the task was to study the work of the joint on glued steel rods on the effect of static and cyclic loads. The requirements are set for the nodes: the strength of the connection should be comparable with the strength of the connected elements; nodes must meet the requirements of universality, in the manufacture of easily assembled at the construction site; nodes of structures for glue and adjacent wood (fiber) during operation perceive both static and cyclic loads.
In modern construction, when the nature of the greenhouse effect provokes significant wind loads in the form of tornadoes, storms and an increase in the pressure of the wind, taking into account the effect of cyclic loading becomes important [2].

3. Research methodology

Studies in this work were conducted on samples of the prisms of the natural sizes of laminated wood, which is along the Central axis was placed the steel rods with a diameter of 14mm class A5 along the fibers on the glue ФРФ50Т [6]. (figure 1).
Static short-term pull-out tests of the samples were carried out on a Shopper tensile testing machine at a scale of 100 kN with a division price of 500 N. The load applied was approximately 1/10 of the load. The sample was suspended from the upper grip of the press using a frame. The prism was mounted by the supporting end on a steel plate of a frame 25 mm thick having a central square hole of 70x70 mm in size. A load was applied to the protruding end of the rod, the sample was centered using a ball joint with a central hole for the passage of reinforcement. When tested for static loads, samples were tested at a speed of 5 mm per minute.
A load of about 500 N was applied and alignment checked. Prism samples were tested for static and cyclic loads according to the scheme of pulling the rod out of the prism with an emphasis on the end face (figure 2).
The samples were tested for the effect of a cyclic load on pulling a rod glued into the wood using the GRM-1 universal hydraulic testing machine using a frame. The machine consists of a control panel and a pulsator.
Some of the samples were tested on a specially manufactured installation with an eccentric type pulsator. The pulsator consisted of a welded frame 1, fixed to the foundation with anchor bolts. A lever 2 is pivotally mounted on the frame.

IOP Conf. Series: Materials Science and Engineering **829** (2020) 012017 doi:10.1088/1757-899X/829/1/012017

The power element is an eccentric type vibrator, which is pivotally mounted on the lever 4 to the lever 2. The vibrator consists of a housing 5 and two gears 6 engaged with eccentrics 7. A flexible shaft from a direct current electric motor is connected to one of the gears, the rotation frequency of which can be adjusted. A cycle counter is connected to it. To create a static load, weights are hung on the vibrator 8, the load is transferred to the sample 11 through the lever 2 and the U-shaped rod 9. A sample installed in the frame 12 is fastened to the U-shaped rod using the collet. The frame is rigidly attached to the power floor face (figure 3).

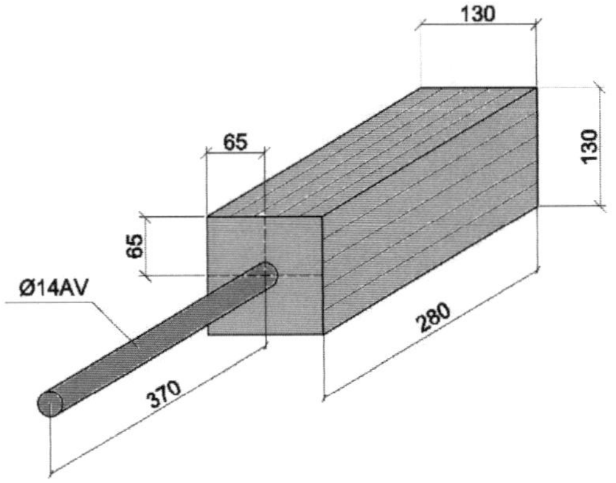

Figure 1. Samples of the prism.

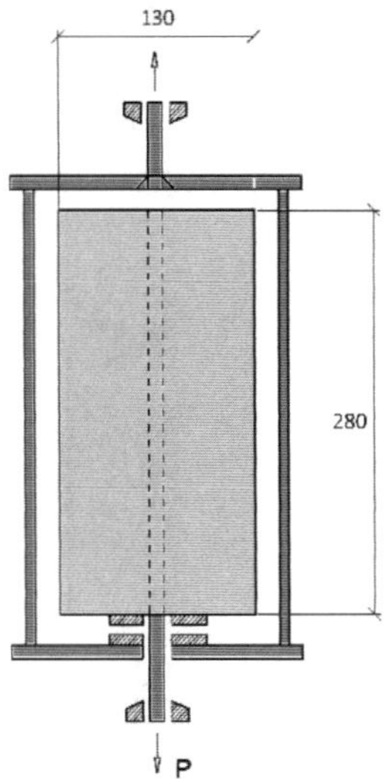

Figure 2. The scheme of testing samples of prisms to pull out the rod according to the scheme with an emphasis on the end.

4. Obtained result

The results of static tests for pulling out a steel reinforcing bar glued wooden prismatic sizes are presented in table 1.

Table 1. Test results.

Breaking load P, [H]	Average value P, [H]	Maximum tangential stress τ i, [MPa]	Static characteristics			
			τ i, [MPa]	Si, [MPa]	γi, [%]	P, [%]
93000	92200	7,56	7,496	4,97	5,4	2,4
95000		7,72				
98000		7,97				
85000		6,91				
90000		7,32				

The nature of the destruction of all samples on wood adjacent to the connection. Together with the core, a piece of wood breaks out of the boundary layer, i.e., destruction occurs as a result of chipping. Cyclic testing of samples for pulling was performed in the range of Pmax and P min. Where Pmax was taken in fractions of a short-term load.

IOP Conf. Series: Materials Science and Engineering **829** (2020) 012017 doi:10.1088/1757-899X/829/1/012017

The frequency of application of the pulsating load was taken 240-300 cycles per minute. The number of cycles during destruction was recorded by a mechanical counter. When the sample was destroyed, the pulsator was automatically turned off.

Tests for cyclic loads were carried out at two cycle asymmetry coefficients $\rho = 0.5$ and $\rho = 0.2$. Only 29 samples were tested: at $\rho = 0.5$ - 20 pieces and $\rho = 0.2$ - 9 pieces.

The maximum stresses at $\rho = 0.5$ were taken 52; 39; 32.5; 26 and 23 kN, which corresponded to the level of load application: 0.81; 0.61; 0.51; 0.41 and, 32 of the short-term strength of the connection equal to 640 MPa.

At $\rho = 0.2$, tests were performed at voltages of 51.0; 32.5; 26.0 and 11.9kN, which corresponds to the level of 0.8; 0.51; 0.41 and 0.28 of σcr.

From each level, 4 to 5 samples were tested with the exception of the lower level, where an increase in the number of test samples did not make sense, since at the previous level the samples were destroyed by fatigue of the reinforcing bar or failure was not observed at a value of 2 2 106 cycles. (see table 2 and 3).

Table 2. Results of tests of samples on endurance at ρ =0.5.

Mark sample	Loading level ($\sigma/\sigma_{кр}$)	Nature of destruction	The number of cycles to failure		Average value (lgN)
			N	**lgN**	
АЦ-1	0.81	on the glue and surrounding wood (fiber)	3000	3,48	3,73
АЦ-2			18000	4,25	
АЦ-3			1500	3,18	
АЦ-4			12000	4,08	
АЦ-5			45000	3,65	
БЦ -1	0.61	on the rod	18300	4,26	4,84
БЦ -2			15000	4,18	
БЦ -3		by glue	445500	5,65	
БЦ -4			180000	5,26	
ВЦ-1	0.51	by glue	747000	5,87	5,88
ВЦ-2		on the rod	813200	5,91	
ВЦ-3			577500	5,76	
ВЦ-4			522000	5,72	
ВЦ-5			1135000	6,05	
ДЦ-1	0.41	on the rod	1372000	6,14	6,14
ДЦ-2			1407000	6,15	
ДЦ-3			823500	5,92	
ДЦ-4			2031200	>6,3	
ГЦ-1	0.32	without destruction	2565500	>6,3	6,3
ГЦ-2			2026200	>6,3	

Table 3. Results of tests of samples on endurance at ρ =0.5.

Mark sample	Loading level ($\sigma/\sigma_{кр}$)	Nature of destruction	The number of cycles to failure		Average value (lgN)
			N	lgN	
КЦ-1	0,8	on the glue and surrounding wood (fiber)	2500	3,4	3,43
КЦ-2			3000	3,47	
ЛЦ-1		by glue	327145	5,51	5,25
ЛЦ-2			75601	4,88	
ЛЦ-3		on the rod	234173	5,37	

МЦ -1	0,61	by glue	368500	5,57	5,6
МЦ -2			421000	5,62	
НЦ -1		without destruction	$2 \cdot 10^6$	>6,3	>6,3
НЦ -2			$2 \cdot 10^6$	>6,3	

According to the results of testing the samples, it was found that the samples were destroyed either by glue or by a metal rod. The table shows that at low levels of cyclic load application, destruction is observed mainly along the reinforcing bar. This is due to the fact that high-strength metal has a low endurance limit. The destruction of the metal rod in the samples occurred at the places of exit from the press grip. At high levels of application of the load of destruction on the contact layer of adhesive-wood, as with a static load.

Based on the calculations for these asymmetry coefficients, endurance coefficients for $\rho = 0.5$ - 0.42 for $\rho = 0.2$ - 0.31. By the method of Zhurkova S.I. [7] predicted durability under the action of cyclic loads for joints reinforcing glued in wood on glue FRF-50T. The relationship between the stress level $\sigma / \sigma cr$ and the number of cycles to failure N in the semi-log coordinates for two asymmetry coefficients is shown in figure 4.

For the asymmetry coefficient $\rho = 0.5$ by the least square method, the dependence is obtained:

$$\sigma / \sigma cr = 1.4 - 0.155 \lg N$$ with the correlation coefficient of the equation $\tau = 0.91$

For $\rho = 0.2$, the dependence $\sigma / \sigma cr = 1.39 - 0.172 \log N$ was obtained for the correlation coefficient of the equation $\tau = 0.84$. Deviations of experimental values from theoretical do not exceed 5%.

5. Conclusion

On the basis of the performed experimental studies on the effect of static and cyclic loads on the connections of the elements of wooden structures on the glued fibers of reinforcing bars with FGF-50t glue, the following conclusions are made: the endurance coefficients of the compound under the influence of cyclic loads at two asymmetry coefficients $\rho = 0.5$ and $\rho = 0.2$ are 0.42 and 0.31, respectively.

The investigated compounds of glued wooden structures are used at facilities in Kazakhstan, where there is a significant wind load. This tennis court SK “Daulet” in Nur-Sultan, sports club “Elkonys” in the recreation area Bayanaul and others.

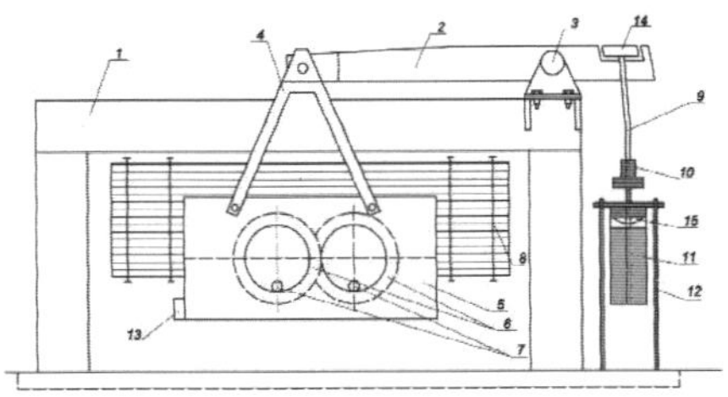

Figure 3. Eccentric type pulsator circuit: 1- pulsator frame; 2- lever; 3 hinge; 4 suspension; 5- vibrator housing; 6-gears; 7- eccentrics; 8-loads; 9-U-shaped traction; 10 collet; 11 sample; 12-frame; 13- limit switch; 14-force meter; 15- hinge.

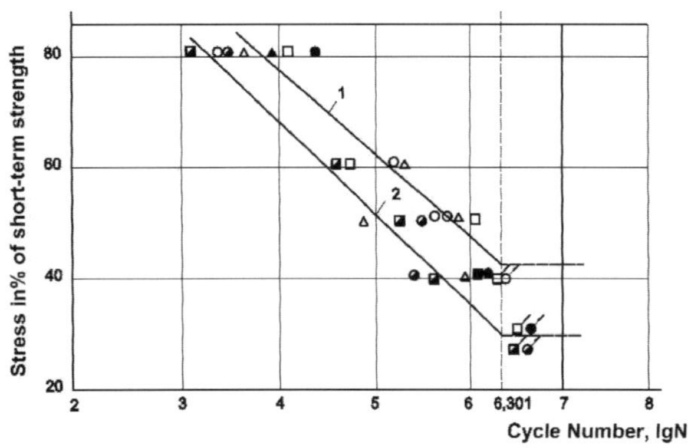

Figure 4. Curves of endurance with $\rho = 0.5$ and $\rho = 0.2$ connection of the reinforcement with wood glue on a modified FGF-50T
(●▲experimental points; - - - direct correlation equation)

6. References

[1] Audrey Farreyre, Jean-bernard journot (2005) Timber trussed arch for long span Master's Thesis in the International Master's Programme in Structural Engineering, Department of Civil and Environmental Engineering Division of Structural Engineering Steel and Timber Structures Chalmers university of technology Göteborg, Sweden.

[2] S. Turkovsky, A. Pogoreltsev, I. Preobrazhenskaya (2013) Central Research Institute of Building Structures. (Russia) Glued wooden structures in modern construction (TsNIISK system). Part 2, Moscow.

[3] Michael Drass, Kay-UweSchober, Michael Kuechler Chalmers (January 2014) with 85 Reads Conference: 13th World Conference on Timber Engineering (WCTE 2014), At Quebec, QC, Canada Cite this publication. University of Technology Advancement of glued-in rods using polymer concrete as composite material Conference Paper (PDF Available)

[4] L Meyer, C Brischke &C R Welzbacher (Published online: 12 Nov 2013) Download citationhttps://doi.org/10.1179/2042645311Y.0000000005. Dynamic and static hardness of wood: method development and comparative studies Pages 5-11 | Received 30 Sep 2010, Accepted 01 Apr 2011

[5] Kenneth Hill, Peter Datin, David O. Prevatt, Ph.D., PE, Kurtis Gurley,Ph.D., Gregory A. Kopp, Ph.D. (2009). 11 Americas Conference of wind engineering. Puerto Rico. A Case for Standardized Dynamic Wind Uplift Pressure Test for Wood Roof Structural Systems. http://www.iawe.org/Proceedings/11ACWE/11ACWE-Hill.pdf

[6] Ospanova Zh.N., Alibekova N.T., Ospanov ES., Aitkhozhin A., Toleubay N.A. (2016) / Bulletin L.N. Gumilyov Eurasian National University – Nur-Sultan: L.N.Gumilyov Eurasian National University, - No. 6 - S.354-358. Heat resistance of glued joints of wood and reinforcing rods

[7] Zhurkov S.N. (1987) // Bulletin of the USSR Academy of Sciences, ser. "Inorganic materials", t.Z, No. 10, p. 1767-1777. Kinetic concept of the strength of solids.

IOP Conf. Series: Materials Science and Engineering **829** (2020) 012018 doi:10.1088/1757-899X/829/1/012018

Properties of Fine-Grained Concrete Mixtures during the Construction of Monolithic Structures

L B Aruova[1*], Z N Ospanova[2], B S Gordienko[3], N T Alibekova[4], Zh E Kalieva[5], Zh I Urkinbaeva[6], M N Nurbaeva[7]

[1]Doctor of Technical Sciences, Professor, Eurasian National University named after L.N. Gumilev, Republic of Kazakhstan
[2,3,4,5]Candidate of Technical Sciences, Associate professor, Eurasian National University named after L.N. Gumilev, Republic of Kazakhstan
[6]Senior Lecturer, Eurasian National University named after L.N. Gumilev, Republic of Kazakhstan
[7]Master of Engineering, Eurasian National University named after L.N. Gumilev, Republic of Kazakhstan

[*] ecoeducation@mail.ru

Abstract. The article presents the results and methods of studying the rheological properties of fine-grained concrete mixtures with various additives, as well as the regression equation obtained by calculations using the apparatus of mathematical analysis, which shows the significance of the parameters included in the process of compaction of the ICZB. These data are necessary for the correct design and selection of sealing mechanisms in order to obtain the design physical and mechanical properties of ICZB. In many regions of the world there is an acute shortage of high-quality raw materials for large aggregate in conventional concrete, which necessitates its transportation over distances from 400 to 1000 km. In this regard, the cost of one cubic meter of coarse aggregate in many regions is 2-3 or more times higher than the cost of sand, the deposits of which are more common. In addition, obtaining the desired fractions of quality aggregates requires artificial addition and fractionation of the resulting granules. The experience of using ICZB for the manufacture of parts and structures at the plants of ZHBK gave positive results, but in mass construction (hydraulic), he did not get spread due to insufficient study of physical and mechanical properties, weak efficiency of existing vibration equipment, insufficient development of new technological equipment for laying and sealing. In this regard, we believe that the research aimed at studying the physical and mechanical properties of ICZB and the further development of technological equipment, methods of laying and sealing for their widespread implementation, determine the relevance of this problem.

1. Introduction
Insufficient study of the physical and mechanical properties of fine-grained concrete mixtures and the inefficiency of the existing vibration equipment to obtain the design physical and mechanical properties of ICZB constrains its use in monolithic construction. Our research in the laboratory and in

experimental areas showed that ICZB has increased frost resistance (this is well explained), the lack of high-quality fillers in it sharply reduces its cost.
A number of researchers dealt with the problem of using ICZB for the manufacture of reinforced concrete structures (fixed formwork), and not for concreting monolithic structures under construction at the construction site. The results of their only indirect bearing on our problem [1, 2].
The main reason that constrains the widespread use of ICZB in monolithic structures lies in the inability to seal them with existing vibration equipment. In the future, it is necessary to continue research on the design and manufacture of the necessary vibration equipment.
The urgency of the task is due to insufficient experience of compaction of fine-grained concrete mixtures in the conditions of the construction site.

2. Purpose of research

In practice, a sufficient amount of experience in the use of ICZB to obtain structures and parts in the factory [8, 11]. However, laboratory, semi-production and production tests of various ISS compositions show that the existing equipment is ineffective in the conditions of the construction site. In this regard, studies aimed at studying the physical, mechanical and rheological properties of ICZB will provide an opportunity to obtain a clear, reasonable idea of the importance of specific parameters for the production of compositions and concretes with the necessary physical and mechanical properties.

3. Method of research

The selection of compositions was carried out according to the well - known technique of Yu.m. Bazhenov with subsequent adjustment, considering that the required properties of ICZB can be obtained with some lack of cement-water paste to fill sand voids.
As additives, widely known air entrainment (SNV), plasticizing (SDB) and slowing down hardening (SP) surfactants were used) [9,10]. The experiments used cement M400 with compression activity of 45.3 MPa, bending-8.66 MPa. The sand was used with a size modulus from 2.0 to 2.8 with a dust particle content of not more than 1 % by weight.

4. Research results.

The results of tests on the compressive strength of different compositions showed THAT the lowest strength of the composition with the air-entraining additive start. This is probably due to the presence of a large amount of air in the concrete (porosity 8-12 %) [4, 6].
All studied compositions withstood 150 cycles of freezing and thawing according to GOST 10060-2012 (MGS) [3]. The best results were obtained for compositions with the addition of SDB in the amount of 0.25 % by weight of cement. In addition, tests were carried out for frost resistance at a temperature of 45 °C. All compositions withstood 15 cycles of thawing and freezing. The decrease in strength was from 1 to 6 %. The test results indicate a high frost resistance of the studied compositions of ICZB.
In parallel with the tests for frost resistance, the modulus of dynamic elasticity of the compositions is determined. The highest dynamic modulus of elasticity, equal to 33.8*103 MPa have compositions with up to-addition of SDB in the amount of 0.25 % by weight of cement.
Shrinkage deformation at an early age (7 days) was slightly higher than that of conventional, but after 180 days they were equal to conventional concrete (0.7 mm/m). MCSB were selected and analyzed for hydraulic structures. Water resistance of all samples from MKZB proved to be of the appropriate grade B-8. Of the studied compositions, MZB with the addition of SDB in the amount of 0.25% by weight of cement has the best properties.
Good frost resistance was obviously provided by the absence of a large filler. A series of experiments and practice have shown that the frost resistance of concrete is provided by the resistance of a large aggregate to alternate freezing and thawing, which is not present in the ICZB. To study the rheological properties of ICZB, a model of sealing equipment was designed and manufactured [7, 5].
The model we have adopted (figure 1) to the greatest extent meets the listed requirements. In its study the following assumptions are made:
- A rigid fine grained concrete mixture is modeled as a viscous elastic material with a modulus of

elasticity E and a damping coefficient b2;

- Plastic deformation does not affect the dynamics of the seal.

As will be shown below, the effectiveness of the vibration profiling on the compacted material depends on its setting, i.e. the set of basic parameters. This setting is carried out by selecting the value of the load to the working body in accordance with the selected frequency and amplitude of the driving force. vibration profiling may be instantaneous, but a preliminary preload of the springs is transformed in the inertial (figure 2).

The vibration lamp is modeled by the inertial mass M1, to which a driving force is applied, created by an unbalanced vibration exciter and the force of the inertial loading of Mo. The layer of compacted fine-grained concrete mixture was modeled by an elastic element and a viscous damper. The parameters of such a shock vibration limiter depend on the elastic modulus E of the concrete mixture, its absorbing properties b2, and the size of the vibration tube. The study of the adopted dynamic system was carried out by methods of mathematical analysis.

For the production task was composed of differential equations of motion of vibration equipment. The basic equation without considering plastic deformations has the form:

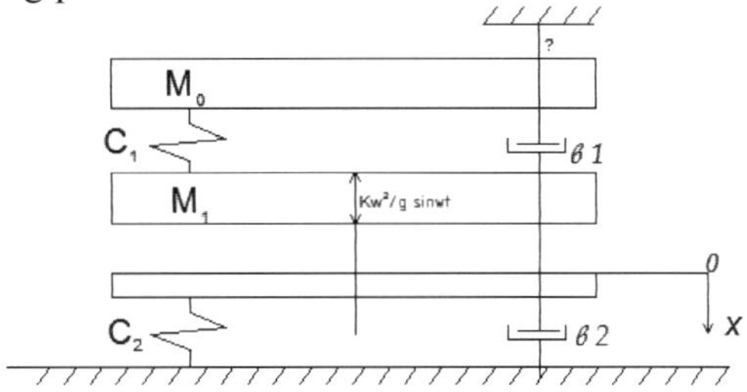

Figure 1. The estimated model for the study of the system "fibro scan – concrete mixture" M0 is the mass of the inertia-free load, b1 and C / 1 are the damping and stiffness coefficients of the springs on which the inertia-free load rests, M1 is the mass of the vibrating parts of the vibration profiling, b2 and C / 2-damping and stiffness coefficients of the limiter (fine-grained concrete mix), Δ – the value of the preload of the springs from the gravity of the inertial preload.

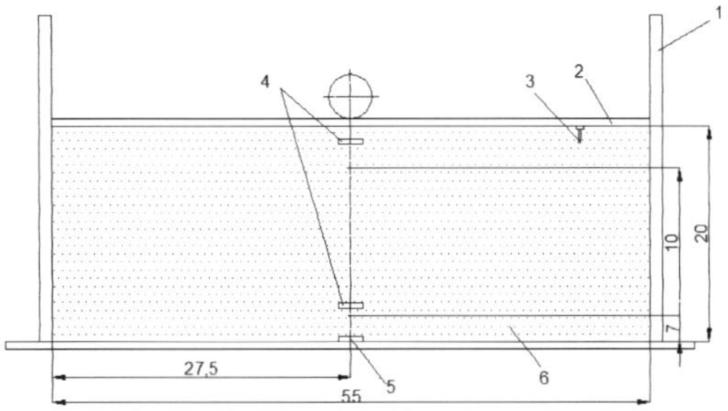

Figure 2. The scheme of placement of sensors. 1-tray formwork, 2-vibration ramming unit, 3-separation sensor, 4 - sensors for measuring accelerations in concrete mix, 5-pressure sensor, 6-fine-grained concrete mix, 7-vibration platform acceleration sensor.

$$M_1 \frac{d^2x}{dt} = C_1 x X - b_1 \frac{dx}{dt} + M_1 * g + c * \Delta + \frac{k\omega^2}{g} * sin\omega t \quad (1)$$

By

$$0 < X < 0, \qquad R\left(X_1 \frac{dx}{dt}\right) > 0 \quad (2)$$

where

$$R\left(X_1\frac{dx}{dt}\right) = C_2 * X + b_2\frac{dx}{dt} \quad (3)$$

$$M_1\frac{d^2x}{dt} = C_1 x X - b_1\frac{dx}{dt} + M_1 * g + c * \Delta + \frac{k\omega^2}{g} * sin\omega t - C_2 * X + b_2\frac{dx}{dt} \quad (4)$$

by

$$X > 0, \qquad R\left(X_1\frac{dx}{dt}\right) < 0 \quad (5)$$

K – static moment of mass of vibration exciter unbalances;
ω – circular frequency of driving force change;
R (X1 $)\frac{dx}{dt}$limiter reaction;
C1, $b$coefficient of stiffness and damping in the suspension load.
X – the coordinate of the vibration lamp; counted from the undamped position of the limiter;
The design model is characterized by 8 parameters: $C_1,\ C_2,\ b_1,\ b_2,\ \Delta, M, K, \omega$. In order to reduce the number of independent parameters we reduce the new variables:

$$Z = \frac{x}{k} * M_1 * g = \frac{x}{a_\infty} \quad (6)$$

$$\tau = \omega t$$

where: $a_\infty = \frac{K}{M_1 * g}$ - amplitude of oscillations of the working body under the action of the driving force $\frac{K*\omega^2}{g} sin\omega t$ in the absence of connections in the air, τ - dimensionless time.
Divide 1 and 4 equations and switching conditions 2 and 5 by the amplitude of the driving force $\frac{K*\omega^2}{g}$.
The equations take a dimensionless form:

$$\frac{d^2z}{d\tau^2} = p - 3_1^2 * Z - \varepsilon_1 * 3_1 * \frac{dz}{d\tau} + \sin\tau \quad (7)$$

by

$$Z < 0 \text{ или } Z > 0, r\left(Z,\frac{dz}{d\tau}\right) > 0 \quad (8)$$

$$\frac{d^2z}{d\tau^2} = p - 3_1^2 * Z - \varepsilon_1 * 3_1 * \frac{dz}{d\tau} + \sin\tau - \varepsilon_2 * 3_2 * \frac{dz}{d\tau} \quad (9)$$

by

$$Z > 0, r\left(Z_1\frac{dz}{d\tau}\right) < 0 \quad (10)$$

where

$$r\left(Z_1\frac{dz}{d\tau}\right) 3_1^2 * Z + \varepsilon_2 * 3_2 * \frac{dz}{d\tau}; p = \frac{(M*g+C*\Delta)*g}{K*\omega^2} \quad (11)$$

From the condition of vibration isolation, we can assume, what $3_1 = 0.2$; $\varepsilon_1 = 0.2$ for all variants of the calculation. In this case the system is characterized by three independent parameters;

IOP Conf. Series: Materials Science and Engineering **829** (2020) 012018 doi:10.1088/1757-899X/829/1/012018

$$З_2^2 = \frac{C_2}{M\omega_2};\ \varepsilon_2 = \frac{b_2}{\sqrt{M*C_2}};p = \frac{(M*g+C*\Delta)*g}{K*\omega^2} \quad (12)$$

In this work the dependencies were investigated p и Z от $З_2$ в области $0.15 \le \varepsilon_2 \le 1.05$, with $З_2$ changed from 0 before 5.
When you search for stable periodic modes of vibration equipment considered such regimes, which provided the maximum magnitude of the stroke speed vibration equipment ceteris paribus. Such regimes were considered the most effective. On the basis of the obtained results, graphs are constructed (figure 3; Fifgure 4). for figure 3) the boundaries of the existence of the most effective regime are determined. Solid lines correspond to the upper limits of the zone of the most effective modes. From the graph (figure 3) it follows that the magnitude of the shock acceleration is approximately in direct proportion to the magnitude of the dimensionless parameter ჳ2.

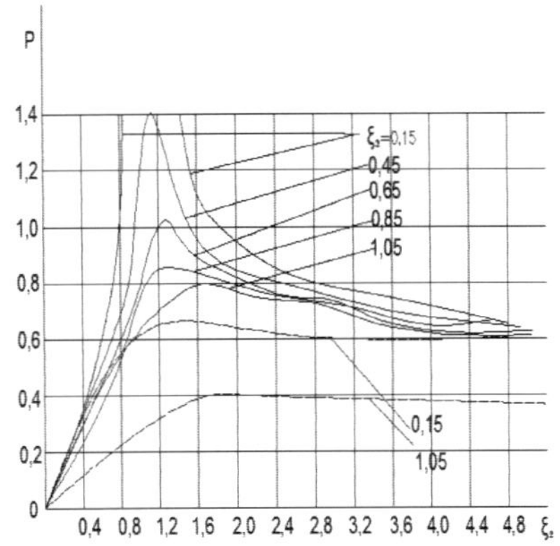

Figure 3. Dependence of the ratio (P) of the gravity of the vibrating parts of the vibration tube to the amplitude of the driving force from the dimensionless parameter ξ2.

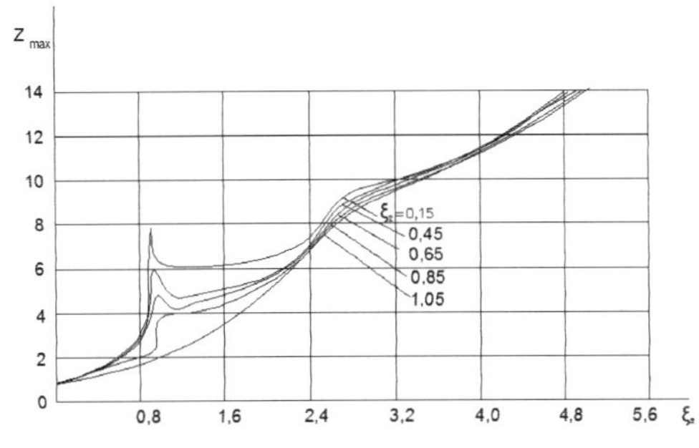

Figure 4. Dependence of shock acceleration Żmax on the vibration lamp from the dimensionless parameter ξ2.

As shown by the study of vibro-viscosity MKZB, using a ball viscometer vibration viscosity of fine-grained concrete mixtures was 5-7 more than for conventional. Increasing the vibro-viscosity leads to a deterioration of the compaction conditions, increased porosity and therefore-to a decrease in strength. A series of experiments on the sealing of the ICSB found that circular vibrators are not effective. After their work in a concrete mix there are not floating glasses. The best results were shown by plane vibration emitters used in massive structures.
To clarify the quantitative and qualitative assessment of the relationship between the thickness of the mixture (h), the stiffness of the mixture (G) and the ratio of the mass of the shock-vibration machine to the amplitude of the driving force $\frac{Q}{P_0}$) in laboratory conditions a series of experiments was carried out using the method of mathematical planning of the experiment.

Compaction of the mixture was carried out by a laboratory shock-vibration machine, which provides the ability to change the amount of load. The purpose of the experiments is to find the optimal parameters of the sealing machine when concreting flattened structures of a certain thickness (h).
In the study, a trial semi-replica of the type was adopted 23-1. The main parameter is the volume mass of the compacted mixture (γ0), which was controlled after each experiment by cores taken from the freshly compacted mixture. After performing all calculations, the regression equation is obtained:

$$\gamma_0 = 2{,}92 - 0{,}37\,\frac{Q}{P_0} + 0{,}066h - 0{,}09Ж$$

When analyzing the regression, it follows that the stiffness of the ICS is within the studied limits (10-20 sec.) has a greater impact on the quality of the seal than the thickness of the mixture layer. The largest coefficient was the ratio of the force of the weight of the shock-vibration machine to the amplitude of the driving force. This confirms the results of computational and theoretical studies.

5. Summary

1. Fine-grained concrete mixes can be applied at concreting of massive (for example, hydraulic engineering) constructions with their compaction by plane vibration emitters.
2. Spread designs it is recommended to seal the sliding vibration lamp with adjustable tightening weight.
3. Fine-grained concrete mixtures are able to replace conventional ones with an acute shortage of large aggregate
4. For achievement of the greatest effect at compaction of fine-grained concrete mixes in each concrete case it is necessary to make adjustment of a vibration lamp on an effective mode.
5. There are areas of effective regimes in dimensionless form (figure 2) and adjust the parameters of vibration lamp need to produce the top borders of these areas. For this ξ2, the effective mode area is bounded by the lower dashed line
6. One of the main characteristics that determine the compaction effect of fine-grained concrete mixtures is the impact acceleration.

6. References

[1] Bazhenov, Y. M. Technology of concrete / Y. M. Bazhenov. - Moscow: Higher school, 2003.
[2] Gladilin, A. V. Econometrics / the A.V. Gladilin, etc. – Reid.: Phoenix, 2011.
[3] GOST 10060-2012 (MGS). Testing of concrete for frost resistance.
[4] GOST 10180-2012 (MGS). Methods for determining the strength of control samples.
[5] GOST 18105-2010 (MGS). Concrete control and evaluation rules.
[6] GOST 22690-2015 (MGS). Determination of strength by mechanical methods of nondestructive testing.
[7] SNiP 5.03-34-2005. Concrete and reinforced concrete structures.
[8] Nanazashvili I. H., Bunkin I. F., Nanazashvili V. I. Building materials and products. Adelant LTD., 2005.
[9] GOST 26633-91 "heavy and fine-grained Concretes".
[10] GOST 24211-91 " Additives in concrete»
[11] Veretennikov Yu. N., Ermakov S. V., Korchagin O. A. fine-Grained concretes for production of ground and sidewalk tiles and stones. Tomsk, TSU Department of "Structures of buildings and structures", 2003.

An Alternative Technique for Accelerated Carbonation of Normal Concrete

Bura Akshay Ramesh[1, a], B. Kondraivendhan[2, b]

[1,2] Applied Mechanics Department, Sardar Vallabhbhai National Institute of Technology, Surat, Ichchhanath, Gujarat-395007, India.

[a] akshayrbura@gmail.com, [b] kondraivendhan78@gmail.com

Abstract. Carbonation is a continuous process in concrete structures and taking place for large period of time. The present study investigates accelerated carbonation effect on normal concrete with the help of Ammonium Bicarbonate (NH_4HCO_3) solution. Normal concrete specimens were exposed to carbonated water curing for different durations after initial 28 days of normal water curing. The effect of carbonated water curing on concrete specimens was studied with the help of compressive strength test and pH profiling test. From the results, it was observed that the compressive strength of normal concrete specimens increased and pH value increased along the depth from the surface with prolonged carbonated water curing.

1. Introduction

Concrete is the most frequently used building material in the civil engineering industry for many years throughout the world. Production of cement is energy intensive and one of the major contributors of CO_2 emission sources in the world. Depending on the production technology, the CO_2 emission ranges from 0.73 to 0.99 t of CO_2 per 1 t of cement. More than one half of this amount is caused by calcinations, which is integral part of the cement production. Looking at the amount of CO_2 production by the concrete industry, it is highly desirable to find a way to reduce the quantity of CO_2 emissions from concrete production. Therefore, there are lots of attempts to use supplementary cementitious materials (SCMs) which can replace at least a part of cement in concrete by more environmental friendly materials. The most common representatives of SCM are industrial by-products such as fly-ash, silica fume or blast furnace slag [1]. Apart from lowering the CO_2 emissions by the construction industry, Carbon Capture and Storage (CCS) techniques have gained importance over the last few decades. A well equipped CCS unit can reduce the CO_2 emissions by 80 – 90 %. Among the various CCS techniques, accelerated carbonation curing (ACC) is considered to be an effective CO_2 sequestration technique that can be adopted for precast concrete units. ACC is different from the carbonation of mature concrete which is considered detrimental to reinforced concrete. The major difference between the two carbonation procedures is its time of application, while the former is intentionally applied within 24 hours of concrete casting; the latter is a normal reaction of cement hydrate products with the atmospheric CO_2 for a prolonged period [2]. Carbonation of cement pastes naturally occurs in atmospheric conditions because CO_2 dissolves in the interstitial solution of the pores of a hardened cement matrix shown by Eq. 1.

$$CO_2 + H_2O \rightarrow H_2CO_3 \rightarrow 2H^+ + CO_3^{2-} \quad (1)$$

In the presence of portlandite, the carbonation reaction occurs as shown in Eq. 2, prior to the formation of calcium silicate hydrate (C-S-H) and Aft and Afm.

$$Ca(OH)_2 + CO_2 \rightarrow CaCO_3 + H_2O \quad (2)$$

Furthermore, the decalcification of C-S-H can occur after the complete carbonation of portlandite, because the stability of C-S-H is highly associated with the alkalinity of the pore solution [3]. The purpose of the present study is to examine the effect of an alternative accelerated carbonation technique using Ammonium Bicarbonate (NH_4HCO_3).

2. Experimental program

In current investigation, experimental study was carried out to analyze an alternative technique for accelerated carbonation of normal concrete. Normal concrete samples were exposed to carbonated water curing for different age periods. The change in compressive strength and pH values were studied after carbonation curing.

3. Materials

Cement used in current study was OPC (Grade 53) confirming to IS: 12269-1987 [4]. Concrete was prepared by using crushed coarse aggregates of size 20 mm MSA (Maximum size of aggregate) and 10 mm MSA. The specific gravity of 20 mm MSA and 10 mm MSA were 2.69 and 2.67 respectively. Locally available natural river sand confirming to Zone II as per IS: 2386-1963 (Part-I) was used as fine aggregate for preparation of concrete mix [5].

4. Mix proportion

In present work, concrete mix design was done in line with DOE method guidelines as per SP23-1982 [6]. Concrete specimens with a constant water-to-cement (W/C) ratio of 0.55 were fabricated. Using W/C ratio and water content of mix proportion, the total binder content was calculated as 309 kg/m^3. Fine aggregate content was taken as 38% of mass of total aggregate. The ratio of 20 mm MSA and 10 mm MSA coarse aggregate were established at 60% and 40% by the total mass of coarse aggregate respectively. The mix proportion of normal concrete is detailed in table 1.

Table 1. Mix proportion of concrete.

Ordinary Portland cement (kg/m^3)	Water content (kg/m^3)	Fine aggregate (kg/m^3)	Course aggregate (kg/m^3)	
			20 mm MSA	10 mm MSA
309	170	724	710	474

5. Preparation of specimens

Concrete mixture was prepared as per the guidelines provided by IS 456-2005 in concrete mixer [7]. Immediately after mixing, the fresh concrete was placed in cube specimens of size 150 mm × 150 mm × 150 mm and vibrated in 3 layers for better compaction. After casting, all the specimens were kept in laboratory environment for 24 hours. All the specimens were demolded after 24 hours and kept in moist curing for the period of 27 days in normal water. Further, concrete specimens were subjected to carbonated water curing after 28 days of initial normal water curing. Carbonation can be induced in concrete specimens by using carbonated water curing as an alternative accelerated carbonation technique. At room temperature, Bakharev et al. (2001) used 0.352 M $NaHCO_3$ solution as carbonated water curing on alkali activated slag concrete [8]. In current study, 0.352 M solution of Ammonium bicarbonate (NH_4HCO_3) was used as carbonated water curing to induce accelerated carbonation in concrete specimens. After every 7 days, carbonated water was renewed to maintain constant carbonated ions (CO_3^{2-}) concentration in curing regime. Compressive strength and pH tests were performed on all the concrete specimens after 6 months of carbonated water curing.

6. Test procedure

6.1. Compressive strength

The test for compressive strength was carried out at the curing ages of 7, 14, 28, 56, 90 and 180 days after initial 28 days of normal water curing. Effect of accelerated carbonation on compressive strength was examined by further curing the concrete specimens in carbonated water beyond initial 28 days of normal water curing. Cube specimens of size 150 mm × 150 mm × 150 mm were used to perform the test in compression testing machine at the rate of 140 $kg/cm^2/min$. For each curing age three samples of concrete specimens were tested and the average of the three was reported as the compressive strength.

6.2. pH profiling of concrete samples

The measurement of pH is essential in studies related to carbonation. Due to carbonation curing, pH drop may occur in pore solution which can be harmful to the rebar preset in reinforced concrete structures. pH of each concrete specimen was measured at discrete layer (0-5, 5-10, 10-15 and upto 25-30 mm from the surface). Concrete powder samples were collected by drilling the concrete specimens at different depth intervals from the surface. For Suspension method of pH determination, concrete powder sample of 4 g was mixed with 80 ml of distilled water in an air tight container. After 24 hours of storage the solution was then tested on a digital pH meter. An apparent pH profile was generated by testing all concrete samples obtained at different depth intervals.

7. Results and discussion

7.1. Compressive strength

The average compressive strength of normal concrete cured in carbonated water at the ages of 7, 14, 28, 56, 90 and 180 days after initial 28 days of normal water curing is shown in figure. 1. It is observed that the compressive strength of normal concrete specimens increased as the duration of carbonated water curing increased. When the specimens were exposed to carbonation curing after initial 28 days of normal water curing, higher compressive strength results were achieved. The carbonation reaction leads to conversion of products from hydration reaction into C-S-H gel or $CaCO_3$. This leads to an increase in the rate of hydration and an accelerated strength gain. Accelerated carbonation curing generates $CaCO_3$ particles in matrix, which fill the pores and reduce the porosity by improving the compressive strength [9]. Thus it can be concluded that prolonged carbonation curing can play an effective role in improving compressive strength with age.

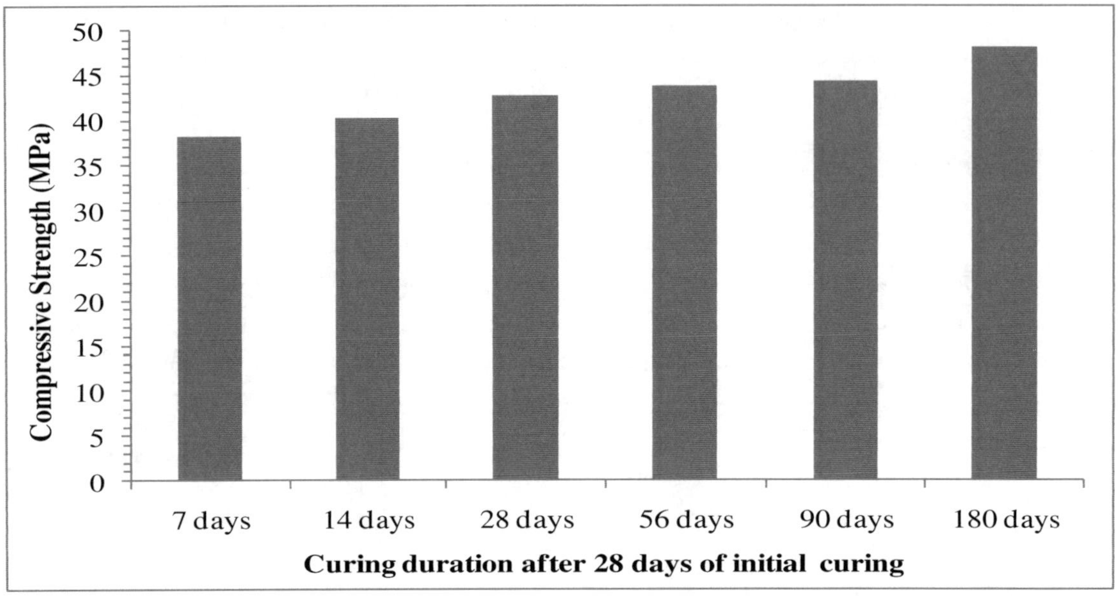

Figure 1. The compressive strength of normal concrete specimens cured in carbonated water with respect to curing duration.

7.2. pH Profiling

Figure 2 present the pH profile of concrete powder samples obtained at various depth intervals from the surface of specimens.

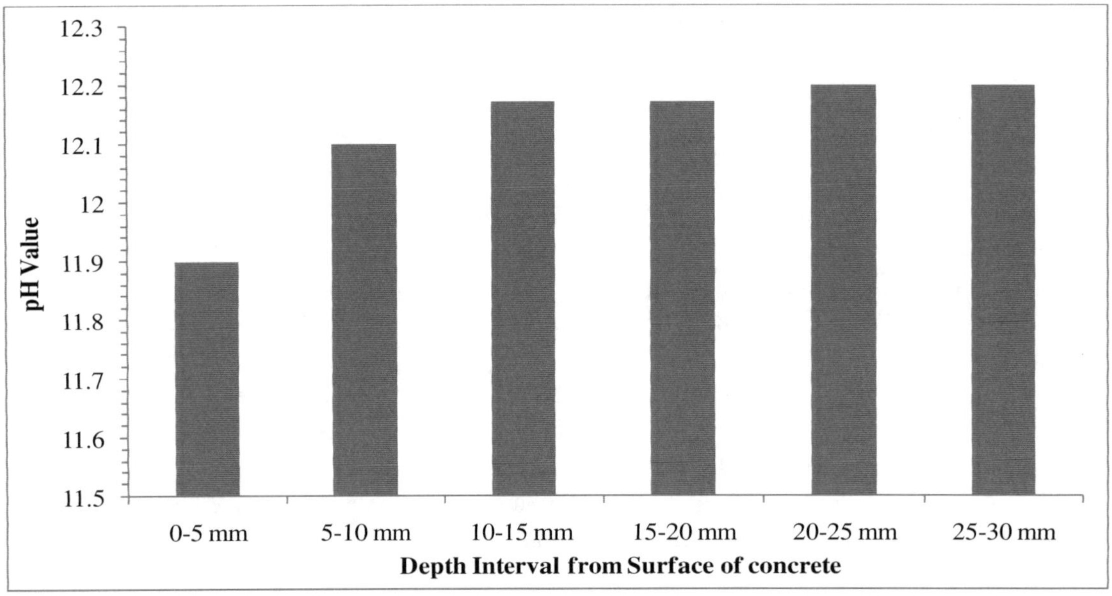

Figure 2. pH profile of the normal concrete samples cured in carbonated water.

From the results it can be observed that pH value was lower at the surface of concrete specimens. After 180 days of carbonated water curing, the pH value of concrete specimens varies from 11.9 to 12.2 between top of the surface to the depth of 30 mm. These reductions in pH value of concrete sample were due to direct exposure of CO_3^{2-} ions to the concrete specimens in accelerated carbonation curing condition. Due to accelerated carbonation curing, pH value of normal concrete reduced due to continuous reduction of calcium hydroxide into $CaCO_3$ as a carbonation reaction product [10].

8. Conclusion

The effect of accelerated carbonation curing on the normal concrete has been evaluated with the help of compressive strength and pH profiling. Based o the experimental results, the following conclusions can be drawn.

1) Accelerated carbonation curing technique increases the compressive strength of normal concrete, due to continuous production of additional C-S-H gel and $CaCO_3$ upon carbonation.

2) The porosity of the concrete specimens decreased upon accelerated carbonation curing, because of densification caused by $CaCO_3$ generation.

3) The pH profiling of normal concrete samples varied depending on curing duration and extend of carbonation from the surface of concrete specimens.

4) The pH value of the concrete samples increased along the depth intervals from the surface due to less carbonation reaction in greater depths.

9. References

[1] Vejmelkova, E., Konakova, D., Kulovana, T., Keppert, M., Zumar, J., Rovnanikova, P., and Cerny, R. (2015). "Engineering properties of concrete containing natural zeolite as supplementary cementitious material: Strength, toughness, durability, and hygrothermal performance." *Cem. Concr. Comp.*, 55, 259-267.

[2] Sharma D. and Goyal S. (2018) "Accelerated carbonation curing of cement mortars containing cement kiln dust: An effective way of CO_2 sequestration and carbon footprint reduction." *Journal of Cleaner Prod.,* 192, 844-854.

[3] Park S. M. and Jang J. G. (2018) "Carbonation-induced weathering effect on cesium retention of cement paste." *Journal of Nuclear Materials 505*, 159-164.

[4] IS: 12269 (1987) "Specification for OPC-53 grade cement." Bureau of Indian Standards, New Delhi.

[5] IS: 2386 (1963) "Methods of test for aggregates for concrete–Part 1: Particle size and shape." Bureau of Indian Standards, New Delhi.

[6] SP23, IS (1982). Handbook on concrete mixes. *Bureau of Indian Standards, New Delhi.*

[7] IS 456 (2000) "Plain and reinforced concrete–code of practice." *Indian Standard, ICS, 91.100. 30, New Delhi.*

[8] Bakharev T., Sanjayan J. G. and Cheng Y. B. (2001) "Resistance of alkali-activated slag concrete to carbonation." *Cem. Concr. Res.*, 31(9), 1277-1283.

[9] Li N., Farzadnia N. and Shi C. (2017) "Microstructural changes in alkali-activated slag mortars induced by accelerated carbonation." *Cem. Concr. Res.*, 100, 214-226.

[10] McPolin D. O., Basheer P. A., Long A. E., Grattan K. T., and Sun T. (2007) "New Test Method to Obtain pH Profiles due to Carbonation of Concretes Containing Supplementary Cementitious Materials." J. Mater. Civ. Eng., 936-946.

IOP Conf. Series: Materials Science and Engineering **829** (2020) 012020 doi:10.1088/1757-899X/829/1/012020

Packing Structure of Binary Particle Compacts with Fibers

A Boribayeva, A Zharbossyn, Z Berkinova, A Yermukhambetova, B Golman*

Department of Chemical and Materials Engineering, School of Engineering and Digital Sciences, Nazarbayev University, Nur-Sultan, Kazakhstan

* boris.golman@nu.edu.kz

Abstract. Fibers have been used to improve the mechanical properties of the asphalt paving mixture. It is known that the enhancement of powder compact mechanical properties is related to the compact packing microstructure. This study focuses on the evaluation of the packing microstructure of powder compacts produced from ternary mixtures of spherical particles and fibers. The discrete element method is employed to generate the compacts of particle mixtures of different compositions under gravity. The compact microstructure is quantitatively characterized by utilizing the developed image analysis technique to approximate the size distribution of voids among particles in X, Y and Z directions. As a result, the denser packing was obtained with a greater fraction of small spherical particles. The inclusion of fibers resulted in the high-density compact with uniform distribution of small size voids.

1. Introduction

Asphalt paving mixture has been in wide use for the production of materials with a strong adhesion effect. The asphalt mixture mainly consists of various sized aggregates. In spite of mechanically strong properties, asphalt pavement is generally affected by a significant traffic load stress leading to irreversible damages and cracking of the surface [1, 2]. Due to the flexibility decrease causing strong deformations [3, 4], the reinforcing of mechanical characteristics of asphalt mixture has become a topical question.

The incorporation of fibers can enhance the consistency of asphalt paving mixture and, thus, significantly improve the mechanical properties of the mixture [5]. Fiber additives have shown good resistance to leakage and drainage through improving the stability of entire mixtures [5, 6]. Moreover, the fibers can be used to lessen various problems with mechanical properties such as dynamic and static stabilities, ductility, and general physical characteristics [7]. It is also known that the aggregate size distribution can have a significant effect on improving the engineering properties of a material [8]. There is still a need in quantitative analysis of the microstructure of powder compact consisting of differently sized particles and fibers, and in research that eventually relates the microstructural parameters to the compact mechanical properties.

Recently, numerical methods have been successfully applied to the investigation of the asphalt compact microstructure due to their cost effectiveness, controlling patterns, and large amounts of data generated. One of the methods to simulate particle packing is a Discrete Element Method (DEM). In early papers, the simulations were conducted with monodisperse spherical particles and only simple interactions between them were taking into account. Lately, DEM has been improved to model more realistic particulates with various characteristics [9, 10]. Yi et al. [9] and Zhu et al. [10] pointed out that it allows accumulating better representation of the dynamical process during compaction in comparison with experimental techniques. The DEM has been recently used to simulate the packing and flow of hot mix asphalt mixtures [11-13].

The objective of this study is to investigate the microstructure of powder compact produced from the ternary mixture of spherical and fiber particles. The DEM will be used to generate the packing structure and the superquadrics approach will be utilized for the creation of spherical and elongated ellipsoidal particles. The compact microstructure will be characterized quantitatively using the size distribution of voids among particles. The image analysis procedure and software will be developed to measure the void size distribution on compact cross-sections in X, Y, and Z directions.

2. Methodology

2.1. Particle generation

In our simulations, the powder compacts consist of the fine and coarse spherical particles with 4 mm and 8 mm in diameters, respectively, and elongated fibers with 12 mm in length. The densities of spherical and fiber particles are 2.257 g/cm^3 [14] and 1.140 g/cm^3 [15], respectively.

The elongated ellipse-shaped particles are used to represent the fibers. The spherical and ellipsoidal particles were created by using a superquadrics approach [16]:

$$f(x,y,z) = \left(\left|\frac{x}{a}\right|^{n_2} + \left|\frac{y}{b}\right|^{n_2} \right)^{\frac{n_1}{n_2}} + \left|\frac{z}{c}\right|^{n_1} - 1 = 0, \quad (1)$$

where a, b and c corresponds to the half-length of the particles, i.e., the radii of fine and coarse particles, and the length of the semi-major ellipse axis, respectively, n_1 and n_2 reflects the sharpness of the particles, and $f(x,y,z)$ is the shape function defined in the local coordinate system. Table 1 illustrates the particle shapes created by the superquadrics builder tool in CFDEM®WORKBENCH software [17, 18].

There are several techniques to increase the time-efficiency of computation, such as controlling the particle shape parameters, size of the particles, managing the overlap between particles, etc. Our preliminary simulations confirmed that the addition of small fibers to the powder mixture results in a significant increase in simulation time owing to the necessity to use a very small time step in DEM simulations. As a consequence, the size of fiber particles has been enlarged to achieve reasonable simulation time.

Table 1. Visualization of spherical and fiber particles created using superquadrics.

		Particles	
		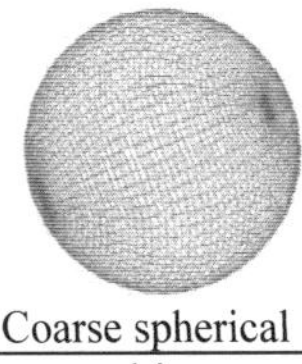	
	Fine spherical	Coarse spherical	Fibers
		Superquadrics parameters	
a, m	0.002	0.004	0.0002875
b, m	0.002	0.004	0.0002875
c, m	0.002	0.004	0.006
n_1	2	2	2
n_2	2	2	2

2.2. DEM simulation setup

The mechanical properties of particles and DEM simulation parameters were taken from Soltanbeigi et al. [16] (see table 2). To study the effect of fibers inclusion on packing structure, DEM simulations have been carried out using the GUI CFDEM®WORKBENCH interface to LIGGGHTS [18] software for four specimens with different mixture composition, as summarized in table 3.

Table 2. The mechanical properties of particles and DEM simulation parameters.

Mechanical properties	
Young's modulus, [Pa]	10^7
Poisson ratio	0.3
Restitution coefficient	0.6
Friction coefficient	0.3
DEM parameters	
Time-step, Δt [s]	$4*10^{-6}$
Gravity, g [m/s^2]	9.81

Table 3. Particle mixture composition for DEM simulations.

Run	Fiber, wt.%	Fine, wt.%	Coarse, wt.%
1	0	63	37
2	1	62	37
3	0	40	60
4	1	40	59

2.3. Analysis of packing structure of powder compact

The packing structure of the ternary mixture of small and large spherical particles and fibers was estimated quantitatively using the concept of void size distribution [19]. The void size distribution is related directly to the packing structure of powder compact [20-22]. The following algorithm was developed to evaluate the size distribution of voids among particles. At first, the three-dimensional image of powder compact simulated using DEM software was visualized in the open-source Paraview software [23]. Then, the Slice Cut function of Paraview was employed to extract the cross-sections of ternary mixture compact taken along with the three axes X, Y, and Z in normal positions. Then, the open-source image analysis software ImageJ [24] was used to convert the color images of cross-sections into the binary images required for subsequent processing. The color images were processed by smoothing images and reducing noise for the clear separation of particles from the void phase. Finally, the distribution of void sizes in a two-dimensional cross-section was measured by randomly placing the void circles of various sizes all over the binary image. The trial was count as a successful one if the circle does not overlap the particle or touch the particle boundary. Then, the probability of the existence of the void of particular size was defined as the ratio of successful trials to the total number of trials. The total number of trials for each circle size was selected as 30,000 based on our preliminary study. The macro program for ImageJ was developed to measure the void size distribution. The examples of cross-section images of powder compact before and after binarization are shown in figure 1. The cross-sections of the compact of a binary mixture of spherical particles are illustrated in figures 1 (a) and 1(b) and the cross-sections of the compact of a ternary mixture of spherical particles and fibers are shown in figures. 1(c) and 1(d). The fibers are clearly seen in figure 1(d), and the particle and fibers outlines in figure 1 (d) are similar to those in figure 1(c) confirming the successful image preparation and binarization procedure.

The microstructure of powder compact was analysed using the cross-sections taken at the same position in the compact in three planes perpendicular to X, Y and Z axes. The area based void size distributions were measured on each cross-section and the distribution parameters were evaluated including the median void size, d_{50}, corresponding to 50% value on the cumulative frequency distribution curve and the distribution sharpness index defined as a ratio d_{90} / d_{10}, where d_{10} and d_{90} are the void diameters corresponding to 10% and 90% of cumulative frequencies, respectively.

3. Results and discussion

3.1. DEM simulation results

Figure 2 illustrates the powder compact generated by DEM simulation and visualized by Paraview for the mixtures of spherical particles (figure 2(a)) and spherical particles and fibers (figure 2(b)) with various fractions of fine and coarse components. The fibers are distributed throughout the compact and the fibers are mainly oriented along with the X and Y directions, as confirmed in figure 2(b).

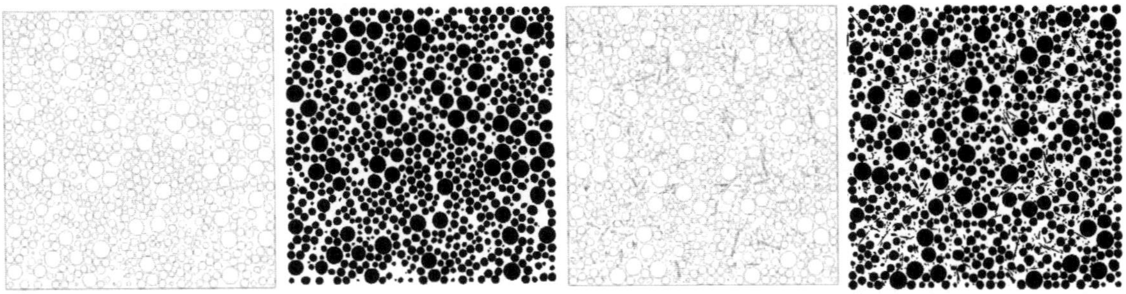

a) b) c) d)

Figure 1. Images of cross-sections normal to Z direction: (a) compact cut from DEM and (b) binary image for Simulation 1, and (c) compact cut from DEM and (d) binary image for Simulation 2.

a) b)

Figure 2. The images of packing structure for (a) Simulation 2 and (b) Simulation 4 in Paraview.

3.2. The microstructure of powder compacts

3.2.1. Binary mixtures of spherical particles. The cumulative frequency void-size distributions measured in X and Y directions on compacts of binary mixtures of spherical particles are shown in figure 3. The median sizes and sharpness indices are summarized in table 4. The ratio of fine and coarse particle sizes was kept constant as 1:2, but the weight fraction of fine particles was varied from 63% (Simulation 1) to 40% (Simulation 3).

The similar trends are observed for packing structures of powder compacts in all directions. The packing densities increases, the size of voids among particles decreases and the voids are more uniformly distributed in the packings of particle mixture with a large amount of fine particles. The median void size decreases from 1.068 mm to 0.914 mm and the sharpness index increases from 0.311 to 0.339 (see table 4, Simulations 1 and 3, X- direction) when the amount of fine particles in the mixture increases from 40% to 63% indicating on the narrow void size distribution with small voids. These trends are more pronounced for packings in X and Y directions.

3.2.2. Ternary mixtures of spherical particles and fibers. To evaluate the effect of fibers on the microstructure of powder compact, fibers (1 wt.%) were added to the mixtures of fine and coarse spherical particles. The compaction of such ternary mixtures under gravity was simulated using DEM, and the void size distributions were measured on obtained powder compacts.

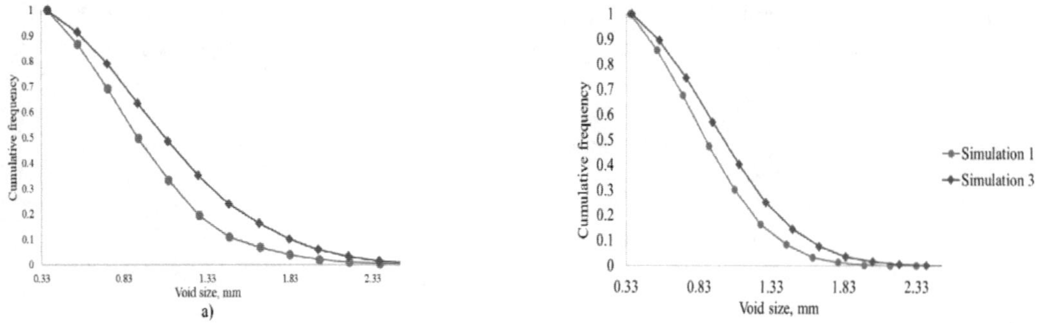

a) b)

Figure 3. Cumulative frequency void-size distributions measured in (a) X, and (b) Y directions for Simulations 1 and 3.

Table 4. Parameters of void size distributions.

Parameters		d_{50}	d_{90}	d_{10}	d_{90}/d_{10}
Simulation 1 (63:37)	X	0.914	0.500	1.475	2.950
	Y	0.864	0.490	1.390	2.837
	Z	0.930	0.485	1.626	3.353
Simulation 2 (1:62:37)	X	0.940	0.528	1.530	3.211
	Y	0.900	0.500	1.457	2.944
	Z	0.795	0.437	1.268	3.109
Simulation 3 (40:60)	X	1.068	0.570	1.830	2.898
	Y	0.998	0.540	1.590	2.914
	Z	0.984	0.533	1.657	2.949
Simulation 4 (1:40:59)	X	0.960	0.540	1.550	2.870
	Y	0.965	0.540	1.515	2.806
	Z	0.870	0.475	1.380	2.905

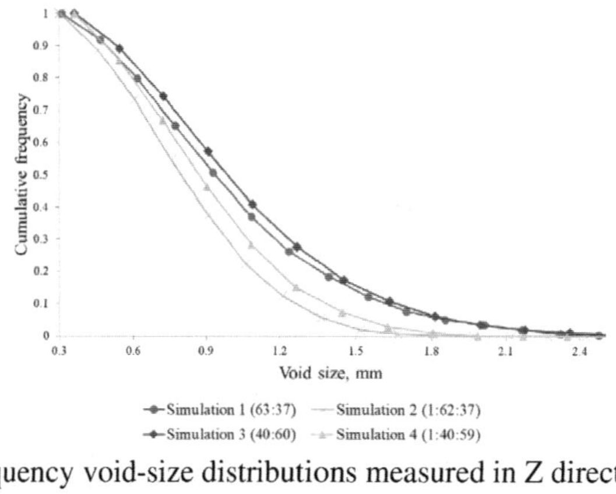

Figure 4. Cumulative frequency void-size distributions measured in Z direction for all samples.

As expected, the addition of fibers has a significant impact on the compact microstructure. The largest differences in void size distributions were observed for cross-sections normal to Z direction due to the preferential orientation of fibers, as shown in figure 4. The size of voids decreases and the distributions become sharper with an addition of fibers for both powder mixtures with low and high amounts of the fine spherical component. The densest packing with the smallest voids (d_{50} = 0.795 mm) and the most uniform size distribution (x_{10} / x_{90} = 0.345) was obtained for the ternary mixture of fibers (1%), and fine (62%) and coarse (37%) spherical particles.

4. Conclusion

The packing microstructure of powder compacts of the ternary mixture of spherical particles and fibers was investigated in the present study. The compacts of particle mixtures of various compositions were simulated using the discrete element method. The fibers were represented by the elongated ellipse-shaped particles. The particles of spherical and ellipsoidal shapes were generated using the superquadrics approach. The packings of binary mixtures of spherical particles and ternary mixtures of spherical particles and fibers were formed under gravity. The compact microstructure was characterized quantitatively by measuring the size distributions of voids among particles in X, Y and Z directions using the developed image analysis procedure. The addition of fibers resulted in denser particle packing with a more uniform distribution of smaller voids. This tendency is more pronounced for the mixture with a larger fraction of fine spherical particles. This analysis allows to evaluate incorporation of fibers on the powder compact microstructure and is useful to design the asphalt paving mixture.

5. References

[1] Akmal N, Usmani A M 1999 *Polym. News* **24** 136
[2] Murali K J, Rajagopal K R 2003 *Appl. Mech. Rev.* **56** 149
[3] Xu Q, Chen H, Prozzi J A 2010 *Constr. Build. Mater.* **24** 2003
[4] Airey G, Collop A, Zoorob S, Elliott C 2008 *Constr. Build. Mater.* **22** 2015
[5] Kaloush K E, Biligiri K P, Zeiada W A, Rodezno M C and Reed J X 2010 *J. Test. Eval.* **38** 400
[6] Hassan H F, Al-Oraimi S and Taha R 2005 *J. Mat. Civ. Eng.* **17** 416
[7] Tapkın S 2008 *Build. Environ.* **43** 1065
[8] Pan T, Tutumluer E and Carpenter S H 2006 *ASCE J. Trans. Eng.* **132** 580
[9] Yi L Y, Dong K J, Zou R P and Yu A B 2011 *Ind. Eng. Chem. Res.* **50** 8773
[10] Zhu H P, Zhou Z Y, Yang R Y and Yu A B 2008 *Chem. Eng. Sci.* **63** 5728
[11] Gong F, Zhou X, You Z, Liu Y and Chen S 2018 *Constr. Build. Mater.* **189** 338
[12] Shashidhar N and Gopalakrishnan K 2006 *Can. J. Civil. Eng.* **33** 945
[13] Pan T, Tutumluer E and Carpenter S H 2006 *ASCE J. Trans. Eng.* **132** 580
[14] Nassar A I, Thom N and Parry T 2016 *Constr. Build. Mater.* **104** 216
[15] Kim M J, Kim S, Yoo D Y and Shin H O 2018 *Constr. Build. Mater.* **178** 233
[16] Soltanbeigi B, Podlozhnyuk A, Papanicolopulos S A, Kloss C, Pirker S and Ooi J Y 2018 *Powder Technol.* **329** 288
[17] Podlozhnyuk A, Pirker S and Kloss Ch 2017 *Comput. Part. Mech.* **4** 101
[18] CFDEM®WORKBENCH, DCS Computing, Linz, Austria (2019)
[19] Kloss C, Goniva C, Hager A, Amberger S and Pirker S 2012 *Prog. Comp. Fluid Dy.* **12** 140
[20] Ohzeki K, Seino K, Kumagai T, Golman B and Shinohara K 2006 *Carbon* **44** 578
[21] Golman B, Seino K, Shinohara K and Ohzeki K 2006 *J. Power Sources* **159** 328
[22] Yermukhambetova A, Berkinova Z and Golman B 2019 *Mater. Today: Proc.* **18** 487
[23] Ayachit U 2015 *The paraview guide: a parallel visualization application*. Kitware, Inc.
[24] Schneider C A, Rasband W S and Eliceiri K W 2012 *Nat. Methods* **9** 671

Acknowledgments
This research was partially supported by the targeted state program BR05236524 "Innovative Materials and Systems for Energy Conversion and Storage" from the Ministry of Education and Science of the Republic of Kazakhstan for 2018-2020. AB and AZ would also like to acknowledge the support of "Young Researchers Alliance" and Nazarbayev University Corporate Fund "Social Development Fund" for grant under their Fostering Research and Innovation Program.

IOP Conf. Series: Materials Science and Engineering **829** (2020) 012021 doi:10.1088/1757-899X/829/1/012021

Shear Strengthening of T-Section RC Beams Using Double-Side Externally Bonded CFRP

Hazem Alfarra[1,2] , Mohamed Maalej[1] and Samer Barakat[1]

[1] Department of Civil & Environmental Engineering, Sustainable Construction Materials and Structural Systems Research Group, University of Sharjah, P.O. Box 27272, Sharjah, UAE

[2] u15200112@sharjah.ac.ae

Abstract. This paper aims at investigating the effect of S-wrap EB-CFRP to shear-strengthen RC T-beams. The experimental program included seven 3.3m-span T-beams that were tested under three-point loading. The specimens were divided into two major groups, including four specimens with no steel stirrups (Group 1), and three specimens with steel stirrups at half of the effective depth (d/2) center to center spacing (Group 2). Five of the seven specimens were strengthened in shear using S-wrap EB-CFRP strips used at d/4 and d/2 spacings. The shear capacity, maximum deflection under the applied load, failure mode, and the strain development in the CFRP strips, steel stirrups, and the main tensile reinforcement were reported in this paper. The overall results show that the addition of EB-CFRP strips to T-beams is an effective method of enhancing the shear capacity of T-Beam in the absence of steel stirrups. With the presence of steel stirrups in the T-beams, the contribution of the EB-CFRP strips in enhancing the beams' shear capacity appears to be small relative to that of steel stirrups.

1. Introduction

Reinforced concrete structures are in danger of deterioration under harsh conditions such as overloading, extreme weather, chemical attacks, and many other factors that will ultimately affect the RC members' structural integrity. Hence, the retrofitting and structural rehabilitation of concrete structures have increasingly become a necessity. One of the widely-used recent technologies in structural enhancements involves the use of externally bonded Fiber Reinforced Polymer (EB-FRP) sheets/strips [1-5]. These technologies have been introduced as alternatives to conventional methods, which come with many limitations, such as being time-consuming, labor-intensive, and costly. The recent technologies, on the other hand, offer significant advantages such as the ease of installation, high durability, high strength to weight ratio, and non-magnetic nature of the FRP [1, 4, 6-9]. The uses of EB-FRP reinforcement may generally be classified as flexural strengthening, shear strengthening, and improving the confinement and ductility of compression members [3, 4, 6, 7].

However, this technique is still relatively new, with limited research that started only less than 25 years ago [3, 5, 8]. This is particularly true for the shear strengthening of T-beams [5], where the presence of the flange in such beams can limit the development of sufficient tensile stress in the EB-FRP to effectively contribute to the shear strength of the member. Moreover, the reported experimental results in the literature showed large scatter when compared to the prediction of the existing design guidelines such as the ACI 440.R2-17.

When strengthening an RC beam in shear, there are various ways in which the EB-FRP systems can be applied. For instance, it can be applied as discrete strips at certain spacing, or as one large continuous band along the member. Moreover, the EB-FRP wrapping scheme could consist of completely wrapping the beam section, using U-shaped wraps, or using double-sided S-wraps, as shown in figure 1.

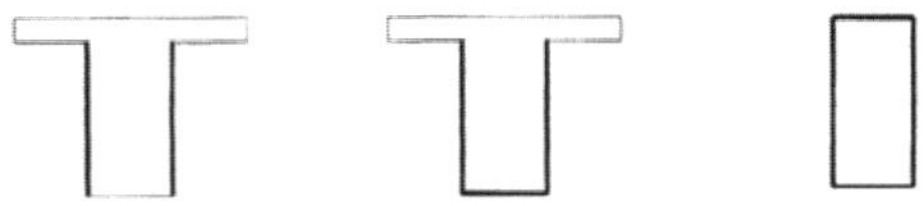

Figure 1. EB-FRP shear strengthening schemes:
S-wrap, U-wrap, and Fully Wrapped (left to right).

Typical failure modes reported in the literature [9-11] for RC beams, strengthened in shear with EB-FRP, include predominantly FRP de-bonding, which usually occurs for FRP U-wrap and S-wrap schemes. This typically happens at a relatively low level of strain in the FRP due to interfacial bond fracture that happens along the FRP-concrete interface, which significantly limits the retrofitting and strengthening performance of these schemes.
The present research focuses on the shear resistance of FRP-strengthened T-beams as they present the closest representation to the real condition of RC beams in practice. As such, fully wrapping the beam would be impractical, and the member shall be strengthened using either U-wraps or S-wraps. In this paper, S-wraps were chosen considering the faster and friendlier installation procedure of this scheme besides the relatively few available researches compared to U-wraps. The research aims to assess the efficiency of the S-wrap scheme in strengthening the T-beams in shear. The shear capacity, maximum deflection under the applied load, failure mode, and the strain development in the CFRP strips, steel stirrups, and the main tensile reinforcement, are used to assess the said efficiency. Furthermore, the interaction between the existing different types of reinforcement (EB-CFRP strips and the steel stirrups) was being under investigation.

2. Experimental investigation

2.1. Material properties
The concrete used to cast the T-beams was a normal weight Ordinary Portland Cement (OPC) concrete with a compressive strength of 47 MPa according to ASTM C39[12]. The details of the steel reinforcements used in the test specimens were as follows:

- 4 #12 mm rebars for the beam flange
- 4 #25 mm rebars for the beam web
- 8 mm rebar for the stirrups

The properties of the reinforcement steel rebars are shown in table 1. Externally-bonded carbon fiber reinforced polymer (EB-CFRP) composites with properties summarized in table 2 are used to strengthen the specimens and were installed according to ACI 440.R2-17 guidelines [13].

Table 1. Steel rebar properties.

Bar Diameter	Yield Strength	Yield Strain [a]
mm	MPa	με
8	592	2960
12	575.4	2877
25	558.6	2793

[a] Yield strain computed considering a modulus of elasticity (E_s) of 200,000 MPa for all bars.

2.2. Test specimens and configuration

The experimental program in the present research included a total of 7 full-scale T- beams that were designed to fail in shear. All beams have similar geometrical properties and longitudinal main reinforcing rebars (ρ_s= 0.0385). The geometric details for the beam section were as follows:

- effective span length (Le): 3300 mm
- effective depth (d): 340 mm
- shear span to effective depth ratio (a/d): 3.2

Further other details are shown in figure 2.

Table 2. CFRP properties.

Properties	Unit	Grade
Fiber Weight	g/cm3	300
Fiber Density	g/cm3	1.8
Fiber Thickness	mm/ply	0.168
Tensile Strength of CFRP	MPa	4,000
Fiber Elasticity Modulus	MPa	230,000
Rupture Strain	με	17,391

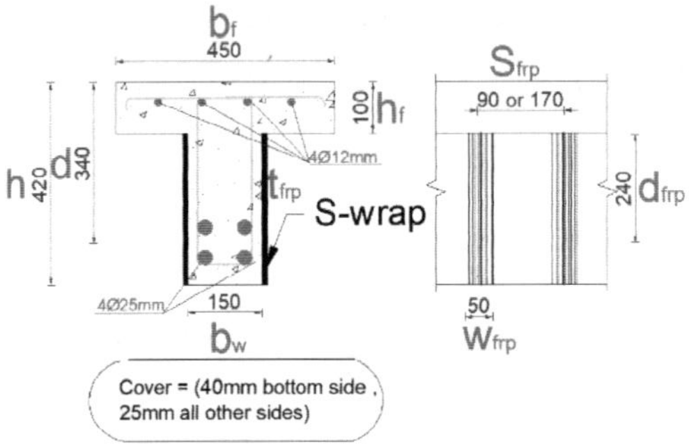

Figure 2. Specimen geometry and reinforcement details (all dimensions are in mm).

To investigate the effect of different variables on the shear behavior of the beams, the specimens were categorized into two main groups, as shown in table 3 and table 4. In these tables, the specimens are named in a way to indicate whether or not they incorporate steel stirrups, EB-CRFP, and the spacing of the steel stirrups and/or EB-CFRP strips. Specimens that are duplicates of other specimens have the letter "D" at the end. As shown in the previous tables, the S-wrap EB-CFRP strips were spaced at 90 mm and 170 mm , these values of spacing have been selected as a quarter and half of the specimens' effective depth (d =340 mm) following the same spacing which was followed in the steel stirrups along the beams of Group 2 (i.e. at 170 mm), in this manner, it will be easier to understand the targeted interaction between the steel stirrups and EB-CFRP strips.

2.2.1. Group 1. Specimens in this group have no steel stirrups and consist of 4 beams labeled as follows:

a) NS-0FC is a control beam without EB-CFRP
b) NS-1F90 is a strengthened beam that incorporates a single layer of EB-CFRP spaced at 90 mm center to center (c/c)
c) NS-1F90D is a duplicate of specimen NS-1F90
d) NS-1F170 is a strengthened beam that incorporates a single layer of EB-CFRP spaced at 170 mm (c/c)

2.2.2. Group 2. Specimens in this group have steel stirrups that are spaced at 170 mm (c/c) and consist of 3 beams labeled as follows:

a) S170-0FC is a control beam without EB-CFRP
b) S170-1F170 is a strengthened beam that incorporates a single layer of EB-CFRP spaced at 170 mm (c/c)
c) S170-1F170D is a duplicate of specimen S170-1F170

Table 3. Specimen reinforcement details.

Beam Name	ρ_v	d_f (mm)	t_{frp} (mm)	w_{frp} (mm)	S_{frp} (mm)	ρ_f
NS-0FC	0	0	0	0	0	0
NS-1F90	0	240	0.166	50	90	0.0012
NS-1F90D	0	240	0.166	50	90	0.0012
NS-2F90D	0	240	0.332	50	90	0.0025
NS-1F170	0	240	0.166	50	170	0.0007
S170-0FC	0.0039	0	0	0	0	0
S170-1F170	0.0039	240	0.166	50	170	0.0007
S170-1F170D	0.0039	240	0.166	50	170	0.0007
S170-1F170D ON	0.0039	240	0.166	50	170	0.0007

Table 4. Test variables under study.

	Beam Name	variables under study
Group 1	NS-0FC NS-1F90 NS-1F90D NS-1F170	• Effect of adding one / two layer of FRP strips at 90 mm on shear behavior • Effect of FRP strips spacing on shear behavior
Group 2	S170-0FC S170-1F170 S170-1F170D	• Effect of adding one layer of FRP strips at 170 mm on shear behavior

2.3. Test setup and instrumentation

Testing of the beam specimens was conducted under displacement control using a three-point loading configuration. The load was applied gradually at a rate of 0.60 mm/minute until beam failure using a 500 kN capacity actuator. The displacement of the beam at three points, including the loading point, was measured using three linear variable differential transducers (LVDTs), as shown in figure 3.

3. Results and discussion

3.1. Failure modes

When loaded, the maximum shear in the beam is expected to develop in the short shear span, which is the subject area of our investigation. As such, most of the shear cracks will develop in this area. For all beams in Group 1, sudden brittle failure was observed. The failure was associated with the development of a major diagonal crack extending from the loading point at the flange and passing through the entire web. This crack was the main reason for the debonding of all EB-CFRP strips. It is worth mentioning that specimen NS-1F90D had an additional crack that started from the loading point and continued horizontally through the flange, although it did not affect the shear capacity of this beam compared to its duplicate counterpart NS-1F90. The failure patterns for all specimens of Group 1 are shown in figure 4.

On the other hand, the failure mechanism of strengthened beams in Group 2 was predominantly ductile with the gradual development of shear cracks as well as the debonding of the EB-CFRP strips. Final failure was accompanied by the development of a nearly-horizontal crack in the flange that started from the loading point and extended beyond the center of the short shear span, in addition to one major diagonal crack, as shown in figure 5. Visually, the beam curvature and the cracking patterns gave sufficient warning about the onset of failure of the beams.

In general, the failure of the EB-CFRP system was by partial debonding, indicating that the EB-CFRP system was not fully utilized up to its maximum capacity. Strain measurements in the CFRP showed that failure took place at CFRP strains well below the CFRP rupture strains shown in table 2.

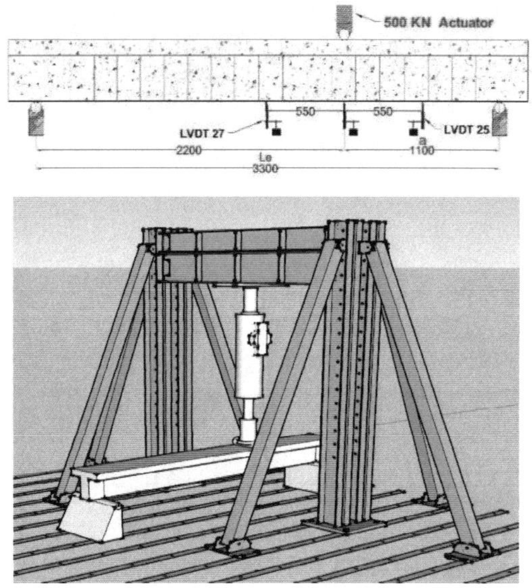

Figure 3. Test setup and instrumentation.

3.2. Load deflection curves

The load-deflection curves of the control and strengthened beams are plotted in figure 6 and figure 7 for Group 1 and 2, respectively. As shown in these figures, the strengthened beams in both groups exhibited an increase in their shear capacities, which was associated with a reduction in the maximum deflection values. For Group 1 beams, one can easily observe from the load-deflection curves the sudden brittle failure. It is worth noting as well that specimen NS-1F90 and its duplicate counterpart (NS-1F90D) showed almost the same load-deflection response. For both specimens, the use of one layer of EB-CFRP increased the beam's shear strength by 41% and reduced its maximum deflection by about 20%, as shown in table 5. For specimen NS-1F170, where the spacing of the EB-CFRP strips was increased from 90 mm to 170 mm, the beam's shear strength increased by only 21%, and its maximum deflection reduced by 17% relative to the control.

For Group 2 beams, which incorporated steel stirrups at 170 mm spacing, the load-deflection responses showed increased ductility in comparison to those of Group 1 specimens. For the former beams (Group 2), the use of one layer of EB-CFRP increased the beam's shear strength by only 9% and reduced its maximum deflection by about 18%. It is worth noting here for specimen S170-1F170(D) that the reduction in the maximum deflection was 34% relative to the control due to premature concrete crashing at the loading point. This result was not considered to be reliable to compute an average reduction in maximum deflection for Group 2 beams.

The above results demonstrate that the use of EB-CFRP in the shear strengthening of T-beams is more effective in the absence of steel stirrups. These results are consistent with findings reported in the literature [14].

Figure 4. Group 1 beams after failure.

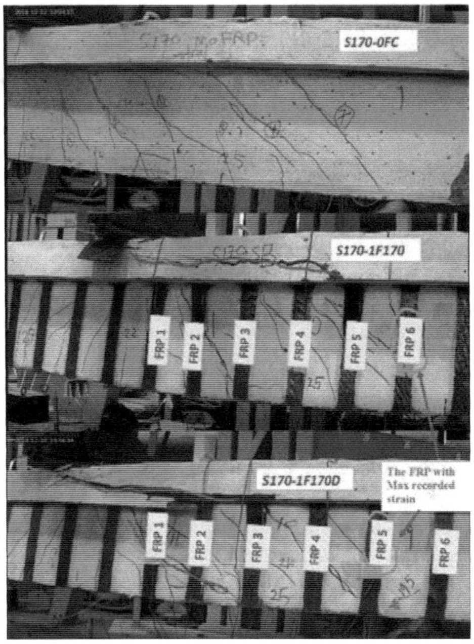

Figure 5. Group 2 beams after failure.

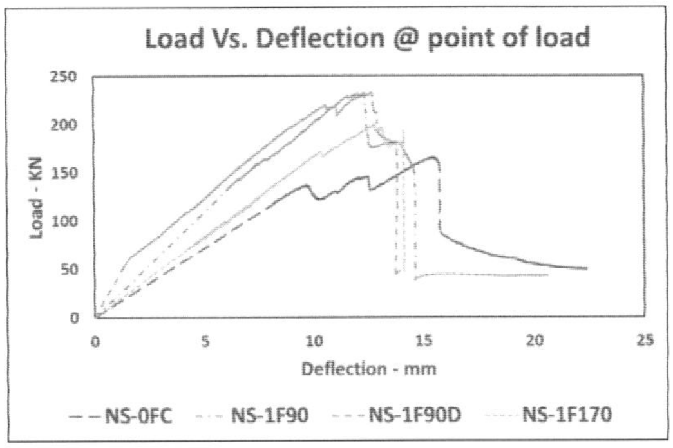

Figure 6. Load-deflection curves of group 1 beams (without steel stirrups).

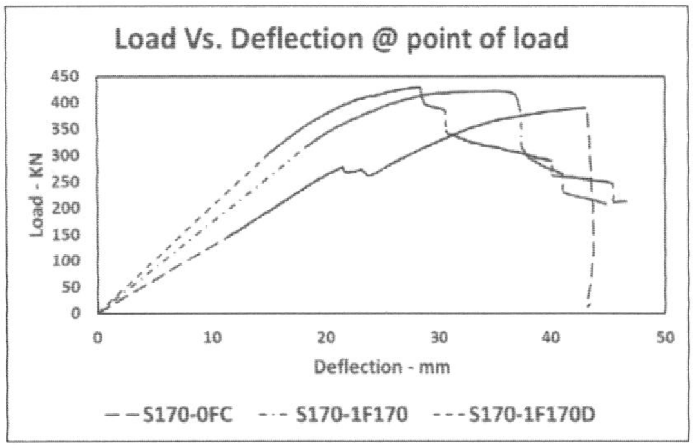

Figure 7. Load-deflection curves of group 2 beams (with steel stirrups).

3.3. Strain response

In an attempt to reach a comprehensive understanding of the interaction between the different types of reinforcement in the T-beams, including the EB-CFRP strips, the steel stirrups, and the main tensile steel rebars, strain gauges were installed in various locations along the short shear span in each type of reinforcement, and the strain measurements were recorded.

Table 5. Summary of load-deflection test results.

	Beam Name	Max Load (KN)	Load Ratio [LR]	Deflection at Max load (mm)	Deflection Ratio [DR]
Group 1	NS-0FC	164.31	1.00	15.52	1.00
	NS-1F90	231.11	1.41	12.32	0.79
	NS-1F90D	231.21	1.41	12.66	0.82
	NS-1F170	198.59	1.21	12.83	0.83
Group 2	S170-0FC	389.26	1.00	42.98	1.00
	S170-1F170	421.36	1.08	35.27	0.82
	S170-1F170D	428.08	1.10	28.25	0.66

For the EB-CFRP strips, strain gauges were numbered as "FRP1" for the strip that lies directly under the applied load, and the number increased until reaching the nearest support, as shown earlier in figure 4 and figure 5. The same procedure was also followed for numbering the strain gauges in the steel stirrups for specimens in Group 2. The maximum measured strain values of all CFRP strips and steel stirrups are shown in table 6, where the strain gauge, which recorded the highest strain for a given specimen, is written between brackets while the missing values during testing are shown in the table as "missing".

For the main tensile steel rebars, only one strain gauge was attached on the reinforcement Rebars. The maximum measured strains in the main tensile steel ranged from about 900 to 1200 $\mu\varepsilon$ among the 4 specimens in Group 1, as shown in table 7 and figure 8. Although the use of EB-CFRP strips has contributed to shear strength enhancements of the T-beams, it did not quite shift the failure mode from shear to flexure, which would have been associated with yielding of the main tensile steel. Furthermore, the maximum measured strains in the EB-CFRP strips ranged from 3700 to 5100 $\mu\varepsilon$ among the three strengthened specimens, as can be seen from table 7 and figure 9.

On the other hand, there was clear yielding of the main tensile steel in Group 2 specimens, which incorporated steel stirrups. The maximum measured strains in the main tensile steel ranged from about 3800 to 4700 $\mu\varepsilon$ among the 3 specimens, as shown in table 7 and figure 8. Strain measurements in the steel stirrups indicate that the latter had also yielded with maximum stirrup strains ranging between 5800 to 7600 $\mu\varepsilon$ among the 3 specimens, as shown in table 7 and figure 10. As for strain measurements in the EB-CFRP strips, they ranged from 2500 to 5800 $\mu\varepsilon$ among the two strengthened specimens, as shown in table 7 and figure 9.

3.4. Comparison of test results with ACI440.2R-17 guidelines

In this section, the contribution of the EB-CFRP (V_f) to the total beam shear strength is calculated from both the experimental results and the ACI440.2R-17 guidelines [13]. From the experimental results, V_f was computed as the difference in shear capacity between the strengthened beams and their control counterparts. For each beam, the shear capacity was computed based on the ultimate load of the beams. The calculated values of V_f from ACI440.2R-17 guidelines did not incorporate any environmental or material strength reduction factors. For each beam, two V_f values were calculated, one based on effective CFRP strain ε_{ef} from the guidelines (Method 1) and another based on experimentally measured ε_{ef} (Method 2) as shown in table 8. A comparison between the experimentally-measured and the predicted V_f values are shown in table 8 for each specimen. Results for duplicate beams are shown in this table as average values.

The results in table 9 indicate that the ACI440.2R-17 guidelines underestimate the EB-CFRP contribution (V_f) to the beam shear strength by about 15 to 20% (Method 1). This seems to be due to the underestimated effective CFRP strain (3565 $\mu\varepsilon$) by the guidelines in comparison to the experimentally-measured values. When the latter is used in computing V_f using the guidelines equations, a nearly-perfect match is observed between the experimentally-measured and the predicted V_f values (Method 2). It is also interesting to note that the experimentally-measured CFRP strain at failure was slightly above the maximum limit (4000 με) allowed by the ACI440.2R-17 guidelines to guard against the loss of aggregate interlock at failure.

The above comparison provides further evidence for the effectiveness of the S-wrap scheme in the shear strengthening of T-beams. The presence of the flange did not seem to negatively affect the capacity of the EB-CFRP strip to develop the necessary tensile strain to contribute to the shear strength of the T-beam.

4. Summary and conclusions

This paper focused on studying the effect of using double-sided (S-wrap) EB-CFRP in the shear strengthening of RC T-beams. Two variables were taken into consideration in the study: the effect of adding one layer of EB-CFRP strips with and without the presence of steel stirrups, and the effect of spacing of the EB-CFRP strips. The experimental program consisted of two groups of specimens, which included seven T-Beams. Two of the specimens were used as control, and five were strengthened with one layer of EB-CFRP. Four of the specimens did not incorporate any steel stirrups (Group 1), and three of the specimens included steel stirrups spaced at d/2 (c/c) (Group 2). The following is a summary of the main findings from the study:

•In the absence of steel stirrups, the addition of one layer of EB-CFRP strips spaced at 90 mm (c/c) resulted in a 41% increase in the shear strength of the T-beams relative to the control. However, this strength increase was associated with about 20% reduction in the deflection capacity. Furthermore, the strength increase was not sufficient to shift the failure mode from shear to flexure. The strengthened specimens failed in a brittle manner without prior warning of impending failure. Strain measurements revealed no yielding of the main tensile reinforcement at failure. By increasing the EB-CFRP strips spacing from 90 mm to 170 mm (c/c), the achieved strength increase was only 21% relative to the control specimen. This strengthening increase was associated with about 17% reduction in maximum deflection.

•In the presence of steel stirrups, the use of EB-CFRP has resulted in about a 9% increase in the beam shear capacity and an 18% decrease in the beams' maximum deflection. The load-displacement response of the specimens showed noticeable ductility in comparison to specimens without steel stirrups. Both the steel stirrups and the main tensile steel showed yielding prior to failure. The latter occurred progressively with an increasing number of shear cracks forming in the web and gradual debonding of the EB-CFRP as the specimen approached the onset of failure.

•According to the above mentioned findings, there was a clear interaction between the steel stirrups and the EB-CFRP strips, as the existence of the steel stirrups spaced at a distance equal to d/2 has inhibited the efficiency of EB-CFRP strips, as they made only a 10% gain in shear strength compared to 41% without the presence of the steel stirrups.

•The predominant mode of failure for all strengthened specimens was debonding of the EB-CFRP strips, triggered by diagonal shear cracks. Final failure was associated with the development of a major diagonal crack extending from the loading point at the flange and passing through the entire web. In some specimens, a nearly-horizontal crack starting from the loading point and extending past the center of the short shear span had also progressively formed together with the diagonal shear cracks.

•The strain measurements in the EB-CFRP strips indicated the effectiveness of the S-wrap scheme in contributing to the shear strength of the T-beams. The presence of the flange did not seem to negatively affect the required development length for the EB-CFRP to effectively contribute to the shear strength of the member. Most of the EB-CFRP strips recorded tensile strains at failure that were greater than the maximum limit (4000 με) allowed by published design guidelines (e.g., ACI 440.2R-17) to guard against the loss of aggregate interlock at failure.

•A comparison between the measured shear strength enhancements (V_f) and those computed as per ACI440.2R-17 guidelines indicated that the latter underestimated the contribution of EB-CFRP to the beam's shear strength by about 15 to 20% among the tested specimens. When the experimentally-measured maximum strain values in the EB-CFRP are used in the ACI440.2R-17 guidelines equation for V_f, a nearly-perfect match is observed between the experimentally-measured and the predicted strength enhancements.

Table 6. Maximum recorded strain values in CFRP strips and steel stirrups (in με).

Group 1				
Beam Name	NS-0FC	NS-1F90	NS-1F90D	NS-1F170
FRP 1	-	416	406	3063
FRP 2	-	652	112	[4288]
FRP 3	-	1638	1033	4201
FRP 4	-	missing	2612	4269
FRP 5	-	4369	3681	3800
FRP 6	-	[5101]	3364	1951
FRP 7	-	2730	[3703]	-
FRP 8	-	2028	3162	-
FRP 9	-	2114	3021	-
FRP 10	-	-31	1661	-
FRP 11	-	617	80	-
FRP 12	-	missing	50	-
Group 2				
Beam Name	S170-0FC	S170-1F170	S170-1F170D	
Stirrup 1	[7485]	5618	[7651]	
Stirrup 2	3329	1641	3751	
Stirrup 3	2609	[5807]	1148	
Stirrup 4	missing	3154	5051	
Stirrup 5	missing	2640	3035	
Stirrup 6	missing	2490	3339	
FRP 1	-	1620	1080	
FRP 2	-	1423	2460	
FRP 3	-	913	2967	
FRP 4	-	2107	2384	
FRP 5	-	1823	[5760]	
FRP 6	-	[2581]	4102	

IOP Conf. Series: Materials Science and Engineering **829** (2020) 012021 doi:10.1088/1757-899X/829/1/012021

Table 7. Summary of recorded maximum strains in EB-CFRP, steel stirrups, and main tensile steel (in $\mu\varepsilon$).

	Beam Name	Max. FRP strain	Max. Stirrup strain	Max. longitudinal reinforcement strain
Group 1	NS-0FC	-	N/A	1022
	NS-1F90	5101	N/A	908
	NS-1F90D	3703	N/A	1219
	NS-1F170	4288	N/A	1152
Group 2	S170-0FC	-	7485	3944
	S170-1F170	2581	5807	3853
	S170-1F170D	5760	7651	4721

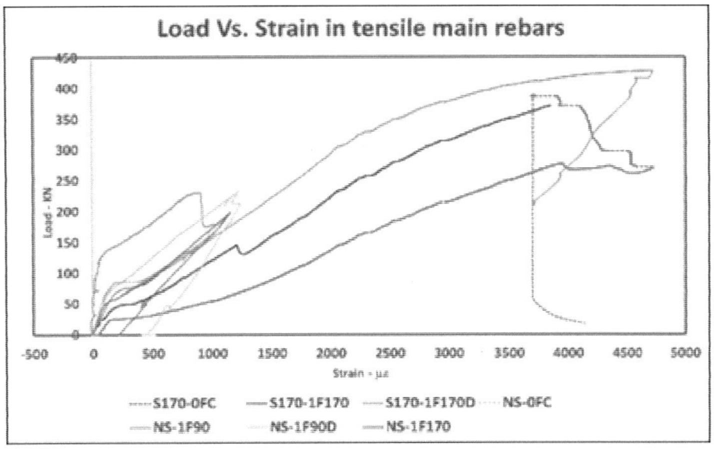

Figure 8. Applied load versus strain in tensile rebars.

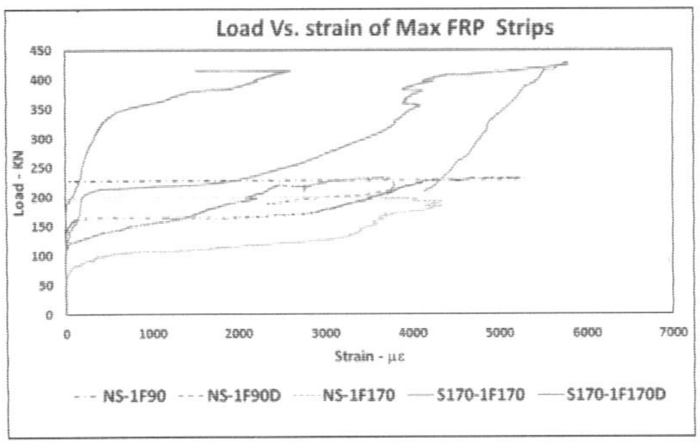

Figure 9. Applied load versus strain in CFRP strips with highest recorded strain.

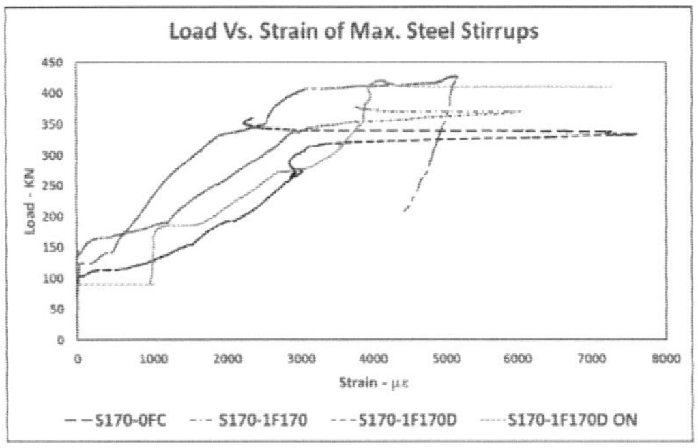

Figure 10. Applied load versus strain in steel stirrup with highest recorded strain.

Table 8. Strain values in CFRP from the ACI440.2R-17 guidelines and based on experimentally measured.

Specimen	Effective CFRP strains as per ACI 440 ($\mu\varepsilon$)	Average CFRP strains from experiment ($\mu\varepsilon$)
NS-1F90/90D	3564.50	4402.00
NS-1F170	3564.50	4288.00
S170-1F170/170D	3564.50	4171.00

Table 9. Summary of recorded maximum strains in EB-CFRP, steel stirrups, and main tensile steel (in $\mu\varepsilon$).

Specimen	Contribution of EB-CFRP to Shear Strength: V_f (KN)			Method 1: V_f^{ACI}/V_f^{Exp}	Method 2: V_f^{ACI}/V_f^{Exp}
	Experimental Results	Method 1: using the effective CFRP strain as per ACI440	Method 2: using average FRP strains from experiment		
NS-1F90/90D	45.60	36.70	45.40	0.80	1.00
NS-1F170	22.80	19.40	23.40	0.85	1.03
S170-1F170/170D	23.40	19.40	22.80	0.83	0.97

5. References

[1] Maalej, M. and K. Leong, Effect of beam size and FRP thickness on interfacial shear stress concentration and failure mode of FRP-strengthened beams. Composites Science and Technology, 2005. 65(7-8): p. 1148-1158.

[2] Perera, R. and A. Ruiz, Design equations for reinforced concrete members strengthened in shear with external FRP reinforcement formulated in an evolutionary multi-objective framework. Composites Part B: Engineering, 2012. 43(2): p. 488-496.

[3] Okeil, A.M., A. Belarbi, and D.A. Kuchma, Reliability of New Shear Design Equations for FRP - strengthened Concrete Bridge Girders. Materials and Infrastructures 1, 2014: p. 101-114.

[4] Baggio, D., K. Soudki, and M. Noel, Strengthening of shear critical RC beams with various FRP systems. Construction and Building Materials, 2014. 66: p. 634-644.

[5] Dirar, S., J. Lees, and C. Morley, Precracked reinforced concrete T-beams repaired in shear with bonded carbon fiber-reinforced polymer sheets. ACI Structural Journal, 2012. 109(2): p. 215-224.

[6] Khalifa, A., et al., Contribution of externally bonded FRP to shear capacity of RC flexural members. Journal of composites for construction, 1998. 2(4): p. 195-202.

[7] Khalifa, A. and A. Nanni, Improving shear capacity of existing RC T-section beams using CFRP composites. Cement and Concrete Composites, 2000. 22(3): p. 165-174.

[8] Chaallal, O., M. Shahawy, and M. Hassan, Performance of reinforced concrete T-girders strengthened in shear with carbon fiber-reinforced polymer fabric. Structural Journal, 2002. 99(3): p. 335-343.

[9] Triantafillou, T.C. and C.P. Antonopoulos, Design of concrete flexural members strengthened in shear with FRP. Journal of composites for construction, 2000. 4(4): p. 198-205.

[10] Bousselham, A. and O. Chaallal, Shear strengthening reinforced concrete beams with fiber-reinforced polymer: Assessment of influencing parameters and required research. Structural Journal, 2004. 101(2): p. 219-227.

[11] Bonacci, J. and M. Maalej, Behavioral trends of RC beams strengthened with externally bonded FRP. Journal of Composites for Construction, 2001. 5(2): p. 102-113.

[12] ASTM C93, Standard test method for compressive strength of cylindrical concrete specimens. American Society for Testing and Materials, Pennsylvania, 2018.
[13] ACI440 Committee, Guide for the design and construction of externally bonded FRP systems for strengthening concrete structures. ACI 440.2 R-17. Farmington Hills: American Concrete Institute, 2017.
[14] Mofidi, A. and O. Chaallal, Effect of steel stirrups on shear resistance gain due to externally bonded fiber-reinforced polymer strips and sheets. ACI Structural Journal, 2014. 111(2): p. 353.

Acknowledgment
This work was partially supported by the University of Sharjah under grant number 18020401100-P and by the Sustainable Construction Materials and Structural Systems (SCMASS) research group. Supports received from M/s FAST BUILDING CONT. CO., and CONMIX LTD during the experimental program are gratefully acknowledged and appreciated.

Effectiveness of Splitter Plate in Suppressing the Flow-Induced Vibration of a Circular Cylinder

Ilya Qurratu'Ain Binti Kamarul'Arifin[1], L K Quen[1*], H S Kang[2,3], L K Tan[1], Nor' Azizi bin Othman[1] and C L Siow[2]

[1]Department of Mechanical Precision Engineering, Malaysia-Japan International Institute of Technology, Universiti Teknologi Malaysia Kuala Lumpur, 54100 Kuala Lumpur, Malaysia
[2]School of Mechanical Engineering, Faculty of Engineering, Universiti Teknologi Malaysia, 81310 Skudai Johor, Malaysia
[3]Marine Technology Center, Institute for Vehicle System and Engineering, Universiti Teknologi Malaysia

*lkquen@utm.my

Abstract. The effectiveness of splitter plate as suppression device in reducing the flow-induced vibration (FIV) of a circular cylinder was conducted experimentally. The parameter used in this experiment is the gap separation between splitter plate and the circular cylinder (G/D). The control element of gap separations are varied from 0.5 to 2.0. The experiments were conducted in the wind tunnel which was free from external wind conditions at Aeronautical and Wind Engineering Laboratory (AEROLAB), UTM Kuala Lumpur. A supporting structure was designed and fabricated with the purpose to allow free vibration induced by wind on the circular cylinder. The results were analysed through the response of amplitude and power spectral density of the circular cylinder with splitter plate and compared with bare circular cylinder. For the gap separations of G/D = 1.0, 1.5 and 2.0, the vibration of the circular cylinder were successfully suppressed. The optimal distance for gap separation was at G/D = 1.5 with suppression effectiveness of 82%. For G/D = 0.5, on the contrary, the vibration of the circular cylinder was amplified. As conclusion, splitter plate can function to suppress as well as amplify the vibration of cylinder, depending on the gap separation.

1. Introduction

Flow over a bluff body such as circular cylinder has raised a remarkable attention in fluid mechanics as it is usually related in many practical situations such as infrastructures and buildings. Maruai et al.[1] stated that bluff bodies were portrayed by the substantial flow that separated and corresponded with the vortex shedding that occurred and produced oscillation when being immersed in fluid flow. Therefore, controlling the vortex shedding from bluff bodies is very crucial to avoid any catastrophic accident.

FIV can cause disastrous failure to a system or structure when the shedding frequency of the vortices behind the wake flow is "lock-in" with the structural natural frequency. The lock-in phenomenon is occurred due to the synchronization between the shedding frequency of the vortices and the structural natural frequency. The failure that caused by FIV is usually costly and requires a lot of concern and effort in order to mend it. The phenomenon of FIV have been experienced in various fields such as civil engineering (tall buildings, chimneys, bridges) [2], aerospace industry [3], power transmission (turbine blades, heat exchangers) [4], oil exploration (offshore platforms) [5] and others. Thus, numerous studies have been conducted in order to modify the fluctuating wake flow of the structure and also to suppress the shedding of the vortices.

Basically, in order to alter the wake behind the structure and reduce the FIV, there are two main methods that can be used which are active and passive control techniques. However, passive control techniques have their advantages where the techniques do not require additional power inputs. As stated by Liu et al. [6], the passive control techniques include modifying the near wake flow of the structure by implementing additional devices in the wake flow such as splitter plates and fairings or by altering the obstacle geometry using strakes.

The studies of passive techniques by using splitter plates still have to be done more especially on its suppression performance. To investigate the reduction of the flow-induced vibration (FIV) in a circular cylinder, splitter plates of various gap separation between splitter plate and circular cylinder (G/D) are tested in the present study to figure out the optimum distance for this suppression device.

2. Methodology

The experiments are conducted at the Aeronautical and Wind Engineering Laboratory (AEROLAB) in UTM Kuala Lumpur, Malaysia. A steel supporting structure is fabricated with the measurements of $0.29m\times0.29m\times0.29m$ in length, width and height, respectively. The circular cylinder used in the experiment is made by polyvinyl chloride (PVC). The selection of the PVC material is due to the light weight. The specifications of the PVC cylinder are shown in table 1.

Table 1. Specifications of the circular cylinder.

Specification	**Value**
length of the cylinder (m)	0.175
Outer diameter (m)	0.022
Wall thickness (m)	0.002
Mass (g)	33

At each end of the circular cylinder, two holes are made in order to attach the coiled spring on the supporting structure. The stiffness of the spring used is in low stiffness which is $29N$. The speeds tested are from 1m/s to 5m/s with interval of 0.5m/s. The suppression effectiveness of splitter plate is investigated based on the gap separations between splitter plate and circular cylinder (G/D). The schematic sketch of circular cylinder with splitter plate can be seen in figure 1 where L is the length of the plate.

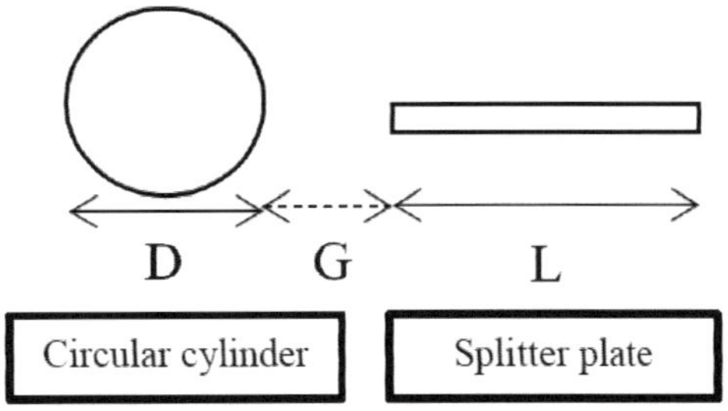

Figure 1. Schematic sketch of circular cylinder with splitter plate.

The values of the parameter that are selected to be used when conducting the experiment are based on table 2.

Table 2. Values of parameter used in the experimental work.

Parameters	Ratio (G/D)
Gap separations between splitter plate and circular cylinder (G/D)	0.5, 1.0, 1.5, 2.0

A video recorder is used to record the motion of the cylinder. The frame per second (FPS) of the recorder is 30, which is adequate to capture the motion of the cylinder. Then, Tracker video analysis and modelling tool software is used to extract the vibration data of the circular cylinder. The video of the circular cylinder that experiencing flow-induced vibration (FIV) needs to be uploaded into the software to be analysed. Before analysing the video, calibration need to be done by setting the coordinate axes and scaling the video using calibration stick. Then, the reference point which can be determined manually is set. After that, the video can be analysed and the amplitude can be obtained.
To get the frequency response, ANSYS software is used by converting the amplitude gained from the Tracker software using Fast-Fourier Transform (FFT). FFT is developed from Discrete Fourier Transform (DFT) with the definition:

$$A(r) = \sum_{k=0}^{N-1} X(s)e^{-i2\pi/N^{rs}} \qquad , r = 0,1,2,\ldots,N-1 \tag{1}$$

where $X(s)$ is the periodic wake and s is the number of samples collected from the original signals.

3. Results and discussions

3.1. Verification of the data

Verification of the experiment is needed before start conducting the experiment. The natural frequency is important as the value is needed for the calculation of reduced velocity, U_R, where U_R = U/ f_nD. U is the flow velocity, f_n is the natural frequency and D is the diameter of the cylinder. The natural frequency can be obtained from the free decay test that is performed by pulling down the cylinder and let it freely vibrates until stagnant position. The time series data from the free decay test is converted into frequency domain using ANSYS software and subsequently, the value of the natural frequency can be obtained. To ensure the validity of the experimental natural frequency, the result is compared with the theoretical value which can be calculated by using the formula:

$$f_n = \frac{\sqrt{k/m}}{2\pi} \tag{2}$$

Where f_n is the natural frequency,k is the stiffness of the system and m represents the mass of the circular cylinder. From table 3, the experimental natural frequency is 8.64 Hz while from the theoretical calculation, the value is 9.44 Hz. The percentage difference between both theoretical and experimental value is 8.47 %. Cecconi et al. [7] stated that the percentage error that usually used as reference has precision of ± 20 % or less. Since the percentage difference of the present study is small, it can be concluded that the experimental rig is reliable and can proceed to the experiment. In the calculation of reduced velocity of the present study, the experimental value of natural frequency, f_n is used instead of the calculated value.

Table 3. Percentage difference on the natural frequency between theoretical and experimental values.

Experimental	Theoretical	Percentage difference (%)
8.64 Hz	9.44 Hz	8.47

3.2. Effect of splitter plate as suppression device

3.2.1. Amplitude response. The ratio form of amplitude can be obtained by dividing the root-mean-square of amplitude with the diameter of the circular cylinder, A/D. Based on figure 2, the maximum amplitude response for a bare cylinder is found to be around 0.16. Lock-in region is started at $U_R \approx 4$ and ended at $U_R \approx 10$. The value of amplitude response at $U_R > 10$ is small as lock-in does not occur. After applying the splitter plate, the amplitude of the cylinder is either becomes larger or smaller based on the gap separation (G/D). In the present study, G/D = 0.5 results the highest amplitude compared to bare circular cylinder and circular cylinder with other gap separation. It is due to the occurrence of violent vibration as the frequency of vortex shedding does not approach the natural frequency of the splitter plate [8]. However, at G/D = 1.0, 1.5 and 2.0, the vibration of the circular cylinder is suppressed as the amplitude ratios are lower than the bare circular cylinder. For G/D = 2.0, the amplitude response seems to be larger compare to G/D = 1.0 and 1.5 as the splitter plate is already too far from the circular cylinder. Hence, the effectiveness of the splitter plate to disrupt the vortex shedding is lesser. The amplitude response of circular cylinder for all gap separation and bare cylinder starting from U_R >13 are in small range as there are no occurrence of FIV.

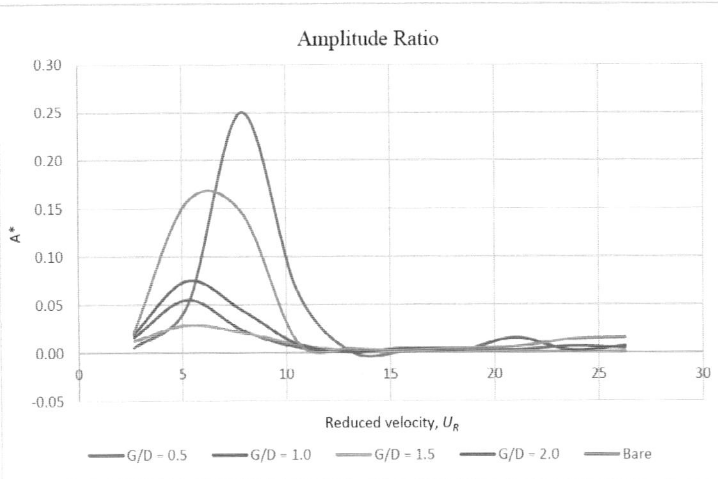

Figure 2. Amplitude ratio of bare cylinder and cylinder fitted with splitter plate of various gap separation.

In order to prove the condition of the lock-in phenomenon, power spectral density for G/D = 0.5 and G/D = 1.5 at wind speed (V) of 1.5 m/s are shown as in figure 3 and figure 4, respectively. It can be seen that the values of power spectral density for G/D = 0.5 are bigger compare to G/D = 1.5 because the lock-in phenomenon is occurred in G/D = 0.5. Moreover, multiple frequency peaks with lower power spectral density are found for G/D = 1.5, while only one significant peak with large power spectral density is found in G/D = 0.5.

3.2.2. Suppression efficiency. The effectiveness of the splitter plate have to be determined to justify the performance of splitter plate based on gap separation. The effectiveness of the splitter plate based on the maximum amplitude can be determined by using the expression as follows:

$$Efficiency\ ratio = \frac{(A/D)_{\max-bare} - (A/D)_{max-splitter\ plate}}{(A/D)_{\max-bare}} \times 100\% \quad (3)$$

Based on table 4 which shows the suppression effectiveness of gap separation, it can be seen the efficiency ratio is the highest at G/D = 1.5 with the value of 81.77%. However, for G/D = 0.5, the efficiency ratio shows a negative value as instead of suppressing the amplitude, the vibration of the cylinder is amplified. For G/D = 1.0 and 2.0, both of the gap separations are able to suppress the vibration of the circular cylinder with moderate effectiveness. For G/D = 2.0, the effectiveness ratio is slightly lower as the splitter plate is too far from the cylinder, hence less disruption on the vortex shedding is experienced by the cylinder.

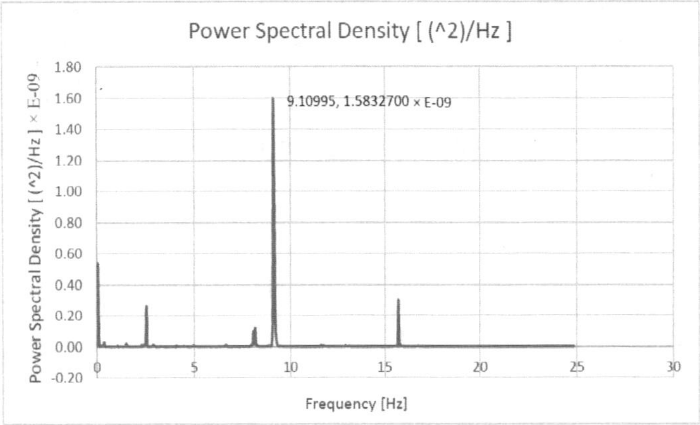

Figure 3. Power spectral density for G/D = 0.5 at V = 1.5m/s.

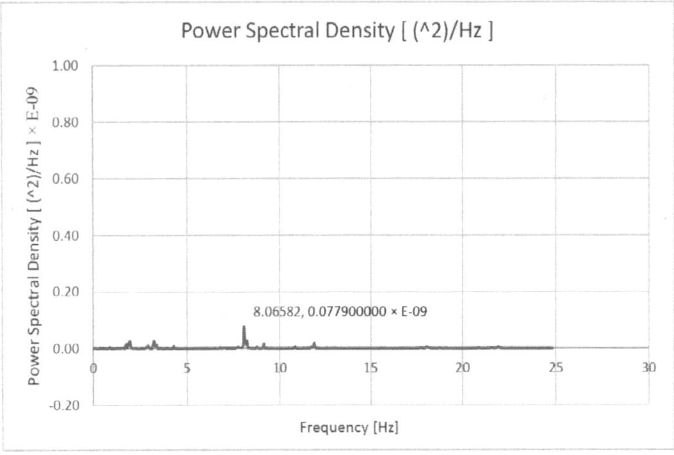

Figure 4. Power spectral density graph for G/D = 1.5 at V = 1.5m/s.

Table 4. Suppression effectiveness based on gap separation between splitter plate and circular cylinder (G/D).

Gap separation between splitter plate and circular cylinder (G/D)	Effectiveness ratio (%)
0.5	-58.23
1.0	65.13
1.5	81.77
2.0	52.59

In conclusion, it is effective to use splitter plate to reduce the FIV of a circular cylinder. However, limitation is found on G/D where the amplitude can be suppressed but also can be amplified. Hence, proper gap selection is required in order to avoid excessive vibration. For this experiment, the most successive gap separation is at G/D = 1.5. Luk et al. [8] conducted an experiment on the reduction of FIV of circular cylinder using air foil as the suppression device. The authors reported that the frequency and strength of the cylinder's vortices dropped when the air foil was located at the distance of 1.5D behind the cylinder. Although the suppression devices involved are dissimilar (airfoil and splitter plate), it is interesting to find out that the outcomes by Luk et al. [8] are similar with the results obtained in this experiment which the optimum gap separation between the circular cylinder and the suppression device is at G/D = 1.5.

4. Conclusions

The FIV phenomenon is occurring in many areas especially in engineering applications. The damages that caused by FIV are very costly. In order to avoid disastrous incident, suppression devices are highly required. Throughout this research, the suppression effect is investigated by using various separation distance between the splitter plate and the circular cylinder, G/D, which are varied from 0.5 to 2.0. The experiments are conducted in a wind tunnel which is free from external wind condition. A supporting structure is designed and fabricated to suspend the circular cylinder so that it can be excited freely depending on the wind speed. Tracker and ANSYS software are used to obtain the data and results for the amplitude and frequency responses. For the gap separations of G/D = 1.0, 1.5 and 2.0, the vibration of the circular cylinder are successfully suppressed. The optimum gap separation to suppress the vibration is at G/D = 1.5, where the reduction percentage is 81.77 %. However, for G/D = 0.5, the vibration of the circular cylinder is amplified as negative reduction percentage is obtained. There are limitations of the gap separation as the vibration can either be suppressed or amplified.

5. References

[1] Maruai, N. M., Ali, M. S. M., Ismail, M. H., & Zaki, S. A. (2018). Flow-induced vibration of a square cylinder and downstream flat plate associated with micro-scale energy harvester. *Journal of Wind Engineering & Industrial Aerodynamics*, *175*, 264–282. https://doi.org/10.1016/j.jweia.2018.01.010

[2] Kawai, H., 1992. Vortex-induced vibration of tall buildings. Journal of Wind Engineering and Industrial Aerodynamics 41, Issues 1-3, pp 117-128. https://doi.org/10.1016/0167-6105(92)90399-U

[3] Sidlof, P., Vlcek, V., Stepan, M., 2016. Experimental investigation of flow-induced vibration of a pitch–plunge NACA 0015 airfoil under deep dynamic stall. Journal of Fluids and Structures 67, pp 46-59. https://doi.org/10.1016/j.jfluidstructs.2016.08.011

[4] Meskell, C., Pellegrino, A. 2019. ortex Shedding Lock-In due to Pitching Oscillation of a Wind Turbine Blade Section at High Angles of Attack. International Journal of Aerospace Engineering 2019. ttps://doi.org/10.1155/2019/6919505

[5] Xu, W.H., Gao, X.F., Du, J., 2012. The prediction on in-line vortex-induced vibration of slender marine structures. Acta Mechanica Sinica 28, Issue 5, pp 1303-1308. DOI 10.1007/s10409-012-0098-3

[6] Liu, K., Deng, J., & Mei, M. (2016). Experimental study on the confined flow over a circular cylinder with a splitter plate. *Flow Measurement and Instrumentation*, *51*, 95–104. https://doi.org/10.1016/j.flowmeasinst.2016.09.002

[7] Cecconi, M., Rhodes, A., Poloniecki, J., Della Rocca, G., & Grounds, R. M. (2009). Bench-to-bedside review: The importance of the precision of the reference technique in method comparison studies – with specific reference to the measurement of cardiac output. *Critical Care*, *13*(1), 201. https://doi.org/10.1186/cc7129

[8] Luk, K. F., So, R. M. C., Kot, S. C., Lau, Y. L., & Leung, R. C. K. (2008). *Airfoil Vibration Due to Upstream Alternating Vortices Generated by a Circular Cylinder*. 79–87. https://doi.org/10.1115/imece2002-33011

Acknowledgments

This research was financially supported by Japan International Cooperation Agency (JICA) research grant (S.K130000.0543.4Y191) and Universiti Teknologi Malaysia matching grant (Q.K130000.3043.01M75).

IOP Conf. Series: Materials Science and Engineering **829** (2020) 012023 doi:10.1088/1757-899X/829/1/012023

Blast Mitigation Effects on Facade System with Projections Utilizing Fluid-Structure-Fluid Interaction

Deepam Patel[1, a] and Atul K Desai[2,b]

[1]Applied Mechanics Department, S. V. National Institute of Technology, Surat, India - 395007
[2]Applied Mechanics Department, S. V. National Institute of Technology, Surat, India – 395007

[a]deepam.patel@gmail.com, [b]atuldesai61@gmail.com

Abstract. Various studies have been conducted on facade system exposed to blast loading have been explored in various studies. Studies conducted are limited to interaction between the fluid and solid domain neglecting the coupling effects of both domains. The impact upon interacting with a free standing will cause the structure to displace its position in the direction of the impact which will decrease the impact upon the structure. At the same time, the impact induced behind the structure will further reduce the effect of impact. This impact generated will marginally counteract the reduction of pressure on the structure. The displacement of the structure as well as shock-wave generated play a major role in reduction of impact on stationary structure. The effects of stress propagation within the structure have been ignored.

1. Introduction

This research paper focuses on reflection of impact wave on the structure leading to formation of Triple point and further exploring the effect of FSI in Blast Wave Mitigation.

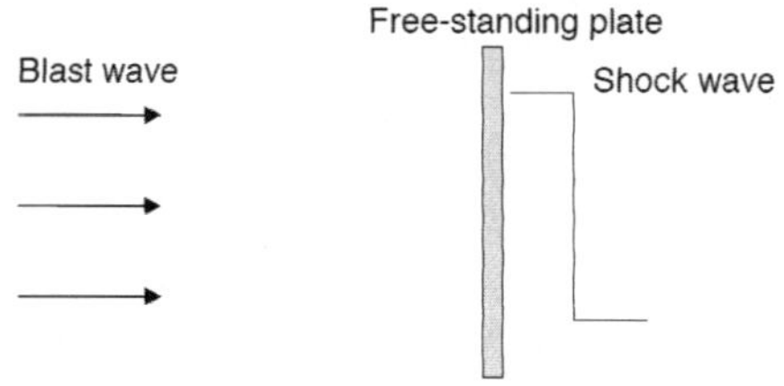

Figure 1. Schematic of a free standing plate subjected to blast wave.

2. Impact wave and an immobilized strcuture

The potential dangers of exposure to explosives can be minimized by carrying out theoretical and numerical simulation. Majority of the research work carried out in past utilize a simplified approach. A numerical model has been created in this work to simulate the effect of an impact wave interacting with an immobilized structure. The effect of stress wave proliferation within in the structure is ignored.

3. Simulation environment

The initial conditions for the flow fields (p_0, T_0, u_0) both sides of the plate are prescribed as the ambient conditions, which are given by

$p_0 = 101.3\text{ kPa},\ T_0 = 298\text{ K},\ u_0 = 0\text{ m/s}$

Initial velocity of the plate, (u_p=0 m/s).

Uniform Blast Wave, $p(t) = p_u$

Where, p_u is a constant overpressure of the blast wave.

Ignoring last negative phase of the Impact Wave,

Exponentially Decaying Blast Wave, $p(t) = p_{is}e^{\frac{-t}{t_i}}$

4. Numerical model validation

4.1. Immobilized Plate

In case of a immobilized plate exposed to exponentially decaying blast wave for simulation of the flow fields, the ratio of Impulsive force transmitted to incident Impulsive force is

$$\frac{I_p}{I_i} = \gamma_R \left(\frac{C_R f_R}{\gamma_R}\right)^{(\beta_s)/(\beta_s+1)} \beta_s^{\beta_s/(1-\beta_s)}$$

where, $C_R = 2\frac{7+4^{p_s}/_{p_0}}{7+^{p_s}/_{p_0}}$, $f_R = \left(\frac{6p_s}{p_0} + 7\right)\sqrt{\frac{(6+C_R)(p_s/p_0)+7}{(p_s/p_0+7)[(1+6C_R)(p_s/p_0)+7][C_R(p_s/p_0)+7]}}$

$\gamma_R = \lim_{\rho_p h_p \to n} \frac{I_p}{I_i} = 8 - 42\frac{p_0}{p_s}\ln\left(1 + \frac{p_s}{7p_0}\right)$, $\beta_s = \frac{\rho_s U_s t_i}{\rho_p h_p}$

Here, p_s = peak overpressure of incident blast wave,

β_s = non-dimensional parameter describing FSI and is inversely proportional to the plate thickness h_p.

ρ_s = density behind the incident shock front, and

U_{sS} = incident blast wave propagation speed.

The results obtained for six sizes of plate thickness and of different blast intensities as shown in figure agree well with the results obtained by Kambouchev et al.With the decrease in plate thickness, impulse ratio decreases whereas for thin plates it becomes very small which suggests the suitability of thin plates for Blast Wave Mitigation.

4.2. Plate moving at a constant velocity

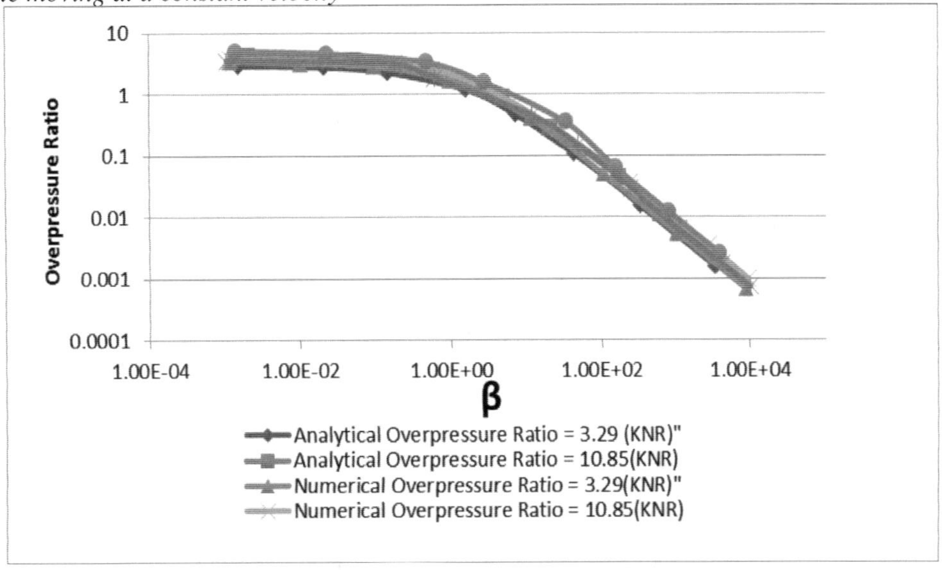

Figure 2. Ratio of Impulse for different blast parameters.
For, low plate velocity ($\frac{p_s}{p_0}$ v/s x, $u_p = 10/s$)

In this case, the shock wave generated behind the plate is induced due to the moving plate in addition to the Blast Wave which can be given by:

$u_p = a_0\left(\frac{(2/\gamma)\frac{p_g}{p_0}}{2\gamma\frac{p_0}{p_g}+(\gamma+1)}\right)^{1/2}$ Where, p_g = overpressure in the back of the shock front, u_p = velocity of the plate.

5. Results and discussion

The numerical simulation for Impact wave between an immobilized plate and a blast wave considering FSI is discussed. Considering the proportion of the reflected impact wave overpressure to the episode impact wave overpressure leads to constraining instances of plates of infinitesimal mass and infinite mass. For air which ranges from 2 to 8.

$$C = \frac{p_r}{p_s} = \frac{(3\gamma-1)\,p_s/p_0 + 4\gamma}{(\gamma-1)\,p_s/p_0 + 2\gamma}$$

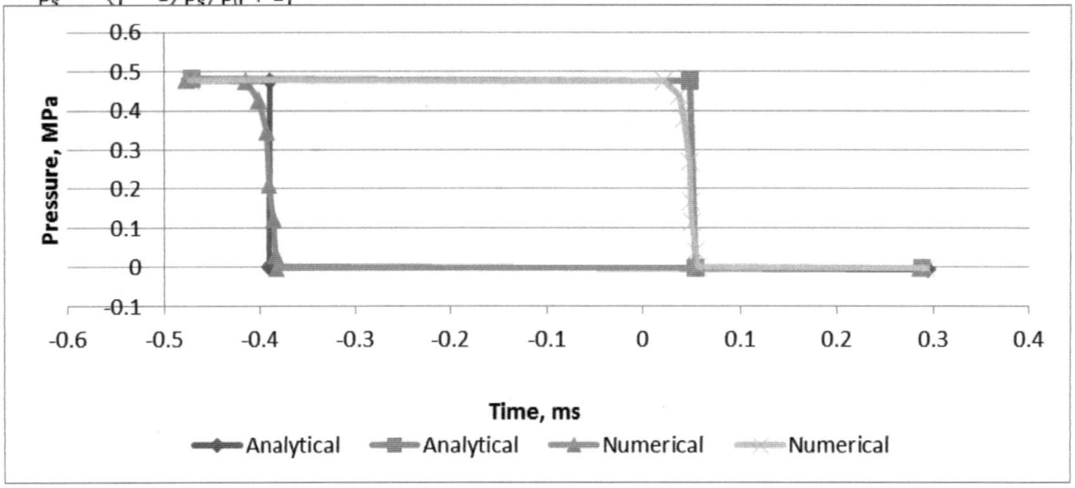

Figure 3. Ratio of Impulse for different blast parameters for different velocities.

(a)

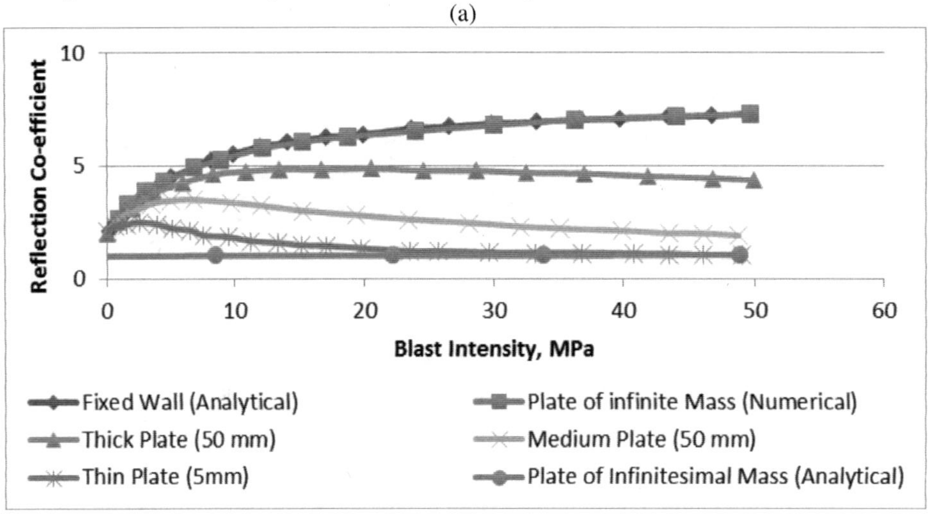

(b)

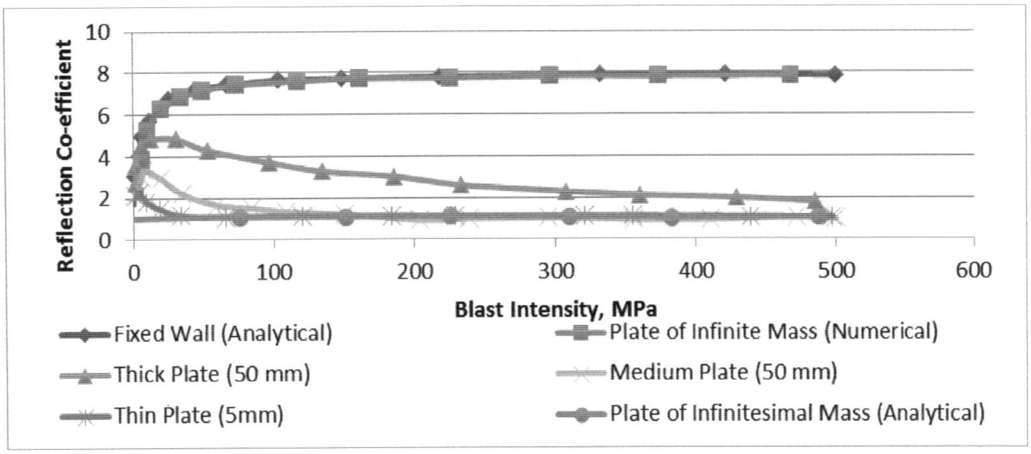

Figure 4. Reflection coefficients for different plate thickness.

Plate thicknesses (0.0005, 0.005, 0.05, 0.5, 5.00, and 50.00 mm) have been considered with the incident impulse (I_i) for low blast intensity of 10.85, with resistance in the back of the plate, is shown in figure 5. An initial Impact wave with 2 MPa Overpressure and decaying time 0.5 ms have been numerically simulated. For infinitesimally light plates, blast waves lead to very high receding velocities which implies the effect of FSI on relatively light plates.Simulations of immobilized plate having thickness 2.50 mm have been subjected to various instances of impulse leading to varied decay time constants.

6. Conclusion

A numerical model of the FSI between an impact wave and an unattached plate has been produced. Considering the blast wave reflection adjoining the structure, for uniform and exponential blast wave, an acceptable validation of numerical and analytical model is achieved. The role of FSI in exponentially decaying blast wave, is limited between two cases. As the intensity of the blast wave increases, the reflection co-efficient asymptotically reached towards unity limited to certain value and thereon reaching a plateau. The reason attributed to this behavior increasing resistance of back of the plate. As the intensity increases, the blast wave impregnates the structure and interacts with the back of the plate leading to higher shock wave in the back of the plate. At the certain thickness and receding velocity conditions, the shock wave generated at the back of the plate is comparable to the shock wave generate in front of the plate leading to diminishing effect of the FSI. Simulations have also lead to conclusion the relationship between ratio of impulse and incident blast intensity. For a particular case, any change in ratio of impulse can significantly affect the incident blast wave.

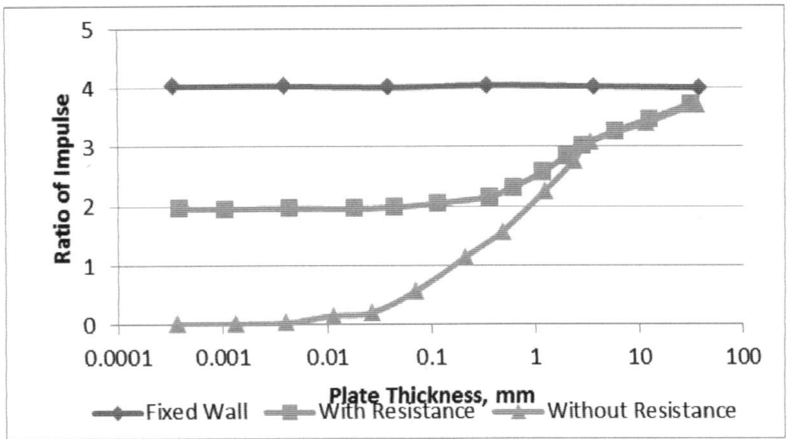

Figure 5. Ratio of impulse for different plate thickness.

7. References

[1] Neuberger, S. Peles and D. Rittel, Scaling the response of circular plates subjected to large and close-range spherical explosions. Part I: Air-blast loading, International Journal of Impact Engineering, Vol. 34, pp. 859-873, 2007.

[2] Vaziri and J. W. Hutchinson, Metal sandwich plates subject to intense air shocks, International Journal of Solids and Structures, Vol. 44, pp. 2021-2035, 2007.

[3] Bram Van Leer, Flux-vector splitting for the Euler equations, Lecture Notes in Physics, Vol. 170, pp. 507-512, 1982.

[4] Y. Tham, Numerical simulation on the interaction of blast waves with a series of aluminum cylinders at near-field, International Journal of Impact Engineering, Vol. 36, pp. 122-131, 2009.

[5] Redekop, Dynamic response of a toroidal shell panel to blast loading, Computers & Structures, Vol. 51, pp. 235-239, 1994.

[6] Espinosa, S. Lee and N. Moldovan, A novel fluid structure interaction experiment to investigate deformation of structural elements subjected to impulsive loading, Experimental Mechanics, Vol. 46, pp. 805-824, 2006.

[7] N. Kambouchev, L. Noels and R. Radovitzky, Numerical simulation of the fluid-structure interaction between air blast waves and free-standing plates, Computers and Structures, Vol. 85, pp. 923-931, 2007.

[8] N. Kambouchev, R. Radovitzky and L. Noels, Fluid-structure interaction effects in the dynamic response of free-standing plates to uniform shock loading, Journal of Applied Mechanics, Vol. 74, No. 5, pp. 1042-1045, 2007.

[9] N. S. Rudrapatna, R. Vaziri and M. D. Olson, Deformation and failure of blast-loaded square plates, International Journal of Impact Engineering, Vol. 22, No. 4, pp. 449-467, 1999.

[10] O. Igra, G. Hu, J. Falcovitz and W. Heilig, Blast wave reflection from wedges, Journal of Fluids Engineering, Vol. 125, No. 3, pp. 510-519, 2003.

[11] Colella, R. E. Ferguson, H. M. Glaz and A. L. Kuhl, Mach reflection from HE driven blast wave, Proceeding of 10th ICDERES, Berkeley, California, Aug. 4-9 1985, AIAA Inc., New York, pp. 388-421, 1986.

[12] W. McCoy and C. T. Sun, Fluid-structure interaction analysis of a thick-section composite cylinder subjected to underwater blast loading, Composite Structures, vol. 37, No. 1, pp. 45-55, 1997.

[13] X.Q. Zhou and H. Hao, Prediction of airblast loads on structures behind a protective barrier, International Journal of Impact Engineering, Vol. 35, pp. 363-375, 2008.

[14] Su, W. Peng, Z. Zhang, G. Gogos, R. Skaggs and B. Cheeseman, Numerical simulation of a novel blast wave mitigation device, International Journal of Impact Engineering, Vol. 35, No. 5, pp. 336-346, 2008.

[15] Z. Xue and J. W. Hutchinson, A comparative study of impulse-resistant metal sandwich plates, International Journal of Impact Engineering, Vol. 30, pp. 1283-1305, 2004.

IOP Conf. Series: Materials Science and Engineering **829** (2020) 012024 doi:10.1088/1757-899X/829/1/012024

Analysis of Infilled Masonry Wall in Steel Frame Subjected to Lateral Loading by Experiment and FEM Software.

Suraj D Bhosale[1], Atul K Desai[2].

[1]Research Scholar, Sardar Vallabhbhai National Institute of Technology, Surat. India.
[2]Professor, Sardar Vallabhbhai National Institute of Technology, Surat. India.

[1]Bhosalesurajd@gmail.com, [2]Atuldesai61@gmail.com

Abstract. This paper describes the behaviour of infilled masonry wall in steel frame subjected to lateral loading. Construction of RCC structure is quite slower than Steel structure, though the cost of steel structure is high but time required for construction is less, moreover for temporary structure steel member and section are use on large scale as it can be reused also. In this study a rectangular steel frame of IS section ISMC 250 is made and lateral loading is applied through jack. Frame is infilled by fly ash units with thin mortar joint. Axial load is also applied on frame simultaneously, during experiment and data is analysed, same full scale model is simulated in FEM tool and result were compare and found that they valid each other, it was observed that in steel frame initially bond between masonry and steel frame section get separated and it try to overturn the masonry wall from toe and debonding at toe initiated but as it is bounded and lock in channel section it stress transfer to heel side of masonry and all stress is concentrated on heel and heel crushing take place.

1. Introduction

Over the last year, the use of confined masonry for low rise structure has increased in developing countries like India. due to confinement its stiffness increases and enhance the performance of the structure in lateral loading or any dynamic loading such earthquake, wind load or in cyclonic loading. Considering they mechanical behaviour of masonry wall as it is intact to column and does not get separated as infilled wall. It helps in carrying lateral and axial load this type of wall is also consider as load bearing wall. In this type of construction wall is cast prior to column and the with the help of tooth action column and wall is made monolithic. As many low-rise buildings made in this way has shown better performance during earthquake in many places.
In tall and slender wall shear crack and diagonal crack failure is pre dominant in lateral loading [4] Number of experimental approaches are adopted to test the masonry wall, such as Prism test, Diagonal compression test of single brick. Generally, prism of 4 to 5 bricks and diagonal compression test of side distance 1 m to 1.2 m. [10]

2. Numerical modelling

Masonry is non linear material, as failure criteria is influencing factor while modelling masonry in FEM tool. There are few approaches made by researchers and they found that some failure criteria satisfy the model failure and validation with experimental results. Concrete Damage Plasticity model is also reliable modelling technic and results are close to real experiment results. In this research micro modelling technics is adopted [12]. Drucker Prager model with expanded XFEM is also showing accurate results to actual existing experiment. [13] There are various modelling strategies adopted as mention below in figure 1. Detail Micro Modelling 2. Simplified Micro Modelling 3. Macro modelling or composite model. In this current study Macro Modelling/ composite modelling is been used to analysis the masonry wall infilled in steel frame.

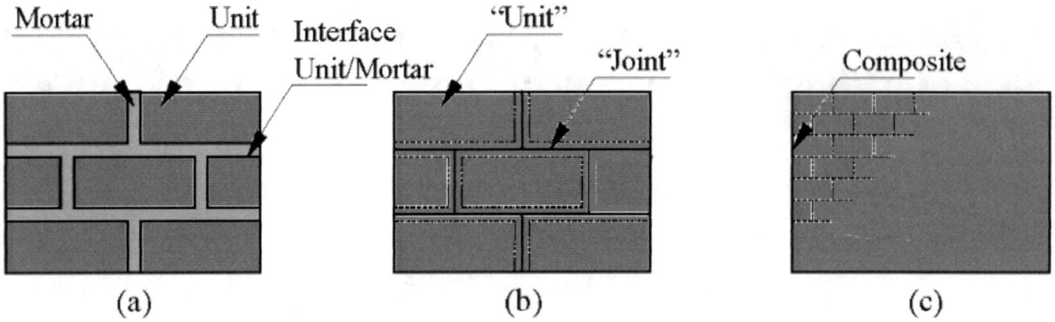

Figure 1. (a) Detail micro modelling (b) Simplified micro modelling (c) macro modelling or composite modelling.

From figure 1 simplest composite model is selected such that the model is homogenous and in which total strength of brick and mortar is considered. This type of modelling is time saving fast and good predictable analysis, where the chances of failure and stress are going to exceed beyond the limit. Can be easily obtained. But this type of modelling is less accurate compare to micro modelling. In this experiment, steel frame of Section ISMC 250 and each unit of size 600mm X 200mm X 100 mm were used as shown in figure 2.

Table 1. Mechanical properties of unit/blocks.

Material	Dry Density	Compressive strength	E (Modulus of Elasticity)	Poisson ratio
Fly Ash Blocks	650 kg/m^3	4.5 N/mm^2	2090 MPa	0.15

3. Experimental setup

Axial load and lateral load are applied through jack which is attach to frame. Frame is made up of Channel section ISMC 250 and bottom two support are hinge support and top two joint are Pinned joint as shown in figure 2. Table 1 shows the mechanical properties of bricks which were used in modelling.

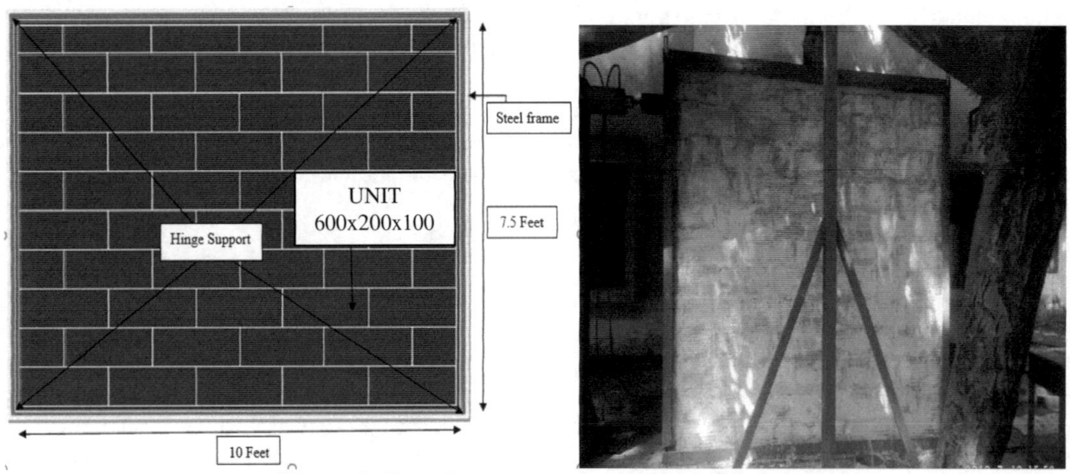

Figure 2. Experiment setup for testing infilled wall.

IOP Conf. Series: Materials Science and Engineering **829** (2020) 012024 doi:10.1088/1757-899X/829/1/012024

Figure 3. Experiment setup bare frame infilled wall.

(a) (b)

Figure 4. (a) Experiment setup and load measuring device. (b) failure at toe and heel due to axial and lateral loading.

The details of experimental setup are shown in figure 3. The arrangement of loading and pressure measuring devices and failure pattern of axial and lateral loading are shown in figure 4. However, figure 5 shows the failure pattern at toe and the crushing at heel due to combine axial and lateral loading.

IOP Conf. Series: Materials Science and Engineering **829** (2020) 012024 doi:10.1088/1757-899X/829/1/012024

(a) (b)

Figure 5. (a) Failure at toe. (b) Failure heel due to axial and lateral loading.

4. FEM Analysis

Figure 6 represents the stress contour after applying axial load, whereas figure 7 shows the stress contour after applying lateral load only. The stress contour of full-scale model developed in FEM software with combine axial and lateral load is shown in figure 8. The obtained FEM results were compared with experimental results, which were in agreement.

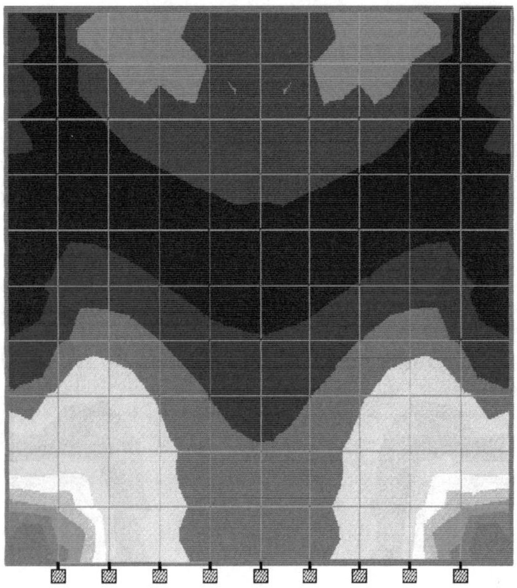

Figure 6. Von mises stresses when axial load is applied.

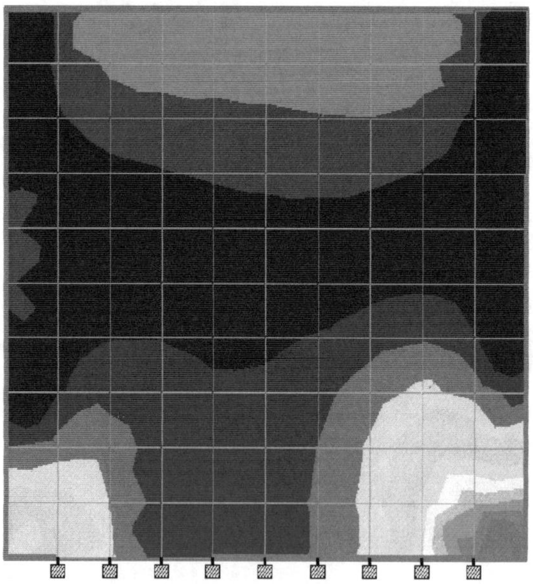

Figure 7. Von mises stresses when lateral loading is applied.

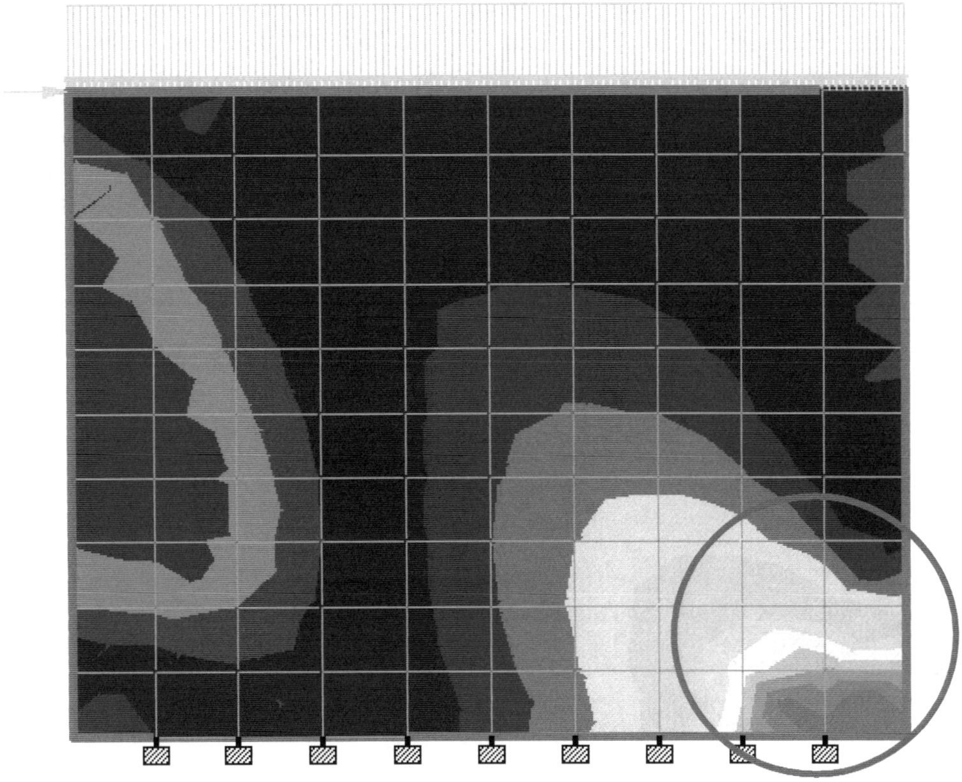

Figure 8. Von mises stresses when lateral load and axial load is applied simultaneously.

5. Conclusion

Significant work has been done on the analysis of masonry wall using FEM Tool. Moreover, experiment is also done to validate the FEM results.

- Above study shows that homogeneous material considering is also a better approach to know the behaviour of masonry wall.
- Using FEM based software can reduce your experimental effort and gives accurate results.
- FEM based Software saves time, and number of variation and combination can be check and analysed.

6. References

[1] Zampilli, M., & Salerno, G. (2019). 5 th International Conference on Structural Health Assessment of Timber Structures The Borbone ' S System : A Multiscale Model For The The Borbone ' S System : A Multiscale Model For The Building Seismic Analysis, (September).

[2] Sousa, L., & Monteiro, R. (2018). Seismic retrofit options for non-structural building partition walls: Impact on loss estimation and cost-benefit analysis. *Engineering Structures*, *161*(November 2017), 8–27.

[3] Silva, L. C., Lourenço, P. B., & Milani, G. (2017). Nonlinear Discrete Homogenized Model for Out-of-Plane Loaded Masonry Walls. *Journal of Structural Engineering*, *143*(9), 04017099. https://doi.org/10.1061/(ASCE)ST.1943-541X.0001831

[4] Rahman, A., & Ueda, T. (2016). In-plane shear performance of masonry walls after strengthening by two different FRPs. *Journal of Composites for Construction*, *20*(5). https://doi.org/10.1061/(ASCE)CC.1943-5614.0000661

[5] Pandey, B. H., & Meguro, K. (2004). Simulation of brick masonry wall behavior under in-plane lateral loading using applied element method. *13th World Conference on Earthquake Engineering, Vancouver, BC, Canada, August*, (1664), 1–6.

[6] Okail, H., Abdelrahman, A., Abdelkhalik, A., & Metwaly, M. (2016). Experimental and analytical investigation of the lateral load response of confined masonry walls. *HBRC Journal*, *12*(1), 33–46. https://doi.org/10.1016/j.hbrcj.2014.09.004

[7] Motovali Emami, S. M., Mohammadi, M., & Lourenço, P. B. (2017). Equivalent Diagonal Strut Method for Masonry Walls in Pinned Connection and Multi-Bay Steel Frames. *Journal of Seismology and Earthquake Engineering*, *19*(4).

[8] Lourenço, P. B., Oliveira, D. V., & Milani, G. (2010). Computational Advances in Masonry Structures: From Mesoscale Modelling to Engineering Application, 1–23.

[9] Fiorino, L., Pali, T., & Landolfo, R. (2018). Out-of-plane seismic design by testing of non-structural lightweight steel drywall partition walls. *Thin-Walled Structures*, *130*(June 2017), 213–230.

[10] Drysdale, R., Hamid, A. A., & Baker, L. R. (1994). Masonry structures: Behaviour and design. Englewood Cliffs, NJ: Prentice Hall.

[11] Bureau of Indian Standards. (2005). IS 2212 - 1991: Code of practice for Brickworks. *IS 2212 - 1991: Code of Practice for Brickworks*.

[12] Bhosale, S. D., & Desai, A. K. (2019). Simulation of Masonry Wall using Concrete Damage Plasticity Model. *International Journal of Innovative Technology and Exploring Engineering*, *8*(9S3), 1241–1244.

[13] Abdulla, K. F., Cunningham, L. S., & Gillie, M. (2017). Simulating masonry wall behaviour using a simplified micro-model approach. *Engineering Structures*, *151*, 349–365.

Acknowledgments
Authors wishing to acknowledge Hilti Company for proper guidance in Anchoring and rebar. Especially Hilti employ Azar and Nehal for on time assessment and guidance, and all my Professor who have advised me and help me in doing experiment work. also, my colleagues who encourage me every time.

IOP Conf. Series: Materials Science and Engineering **829** (2020) 012025 doi:10.1088/1757-899X/829/1/012025

Research on the Influence of Variable Stiffness Levelling Piles on Differential Settlement of High-rise Buildings

Yufeng Yang [1], **Tengjun Gan**[1]

[1] school of Civil Engineering , Chongqing University, China

[1] Ezreal_Yang@163.com

Abstract: Due to the differential settlement of the piled raft foundation of high-rise buildings and super high-rise buildings, secondary stresses will be generated in the superstructure, and variable stiffness leveling can solve this problem. However, different methods of laying piles may cause large differences in the amount of piles used and the bearing capacity of the foundation. In order to explore the effects of different combinations on differential settlement and internal forces, this article studies the upper structure and foundation as a whole, using finite element software ABAQUS, calculation and analysis of piled raft foundations under self-weight and lateral loads. The effects of different pile lengths, different pile diameters, dense piles, and different pile lengths and different pile diameters on the different settlements and internal forces are analyzed. The research found that the variable pile length form has the best effect on the control of differential settlement, followed by the variable pile pitch type, and the effect of the variable pile diameter, the combination of variable pile length and variable pile diameter is poor. The research results can guide the design unit's design of pile distribution, and maximize the economy and safety.

1. Preface

Forty years since the reform and opening up, China 's economy has continued to develop and the process of urbanization has accelerated. Tall and super high-rise buildings can use land efficiently, enhance the image of the city, and promote economic development, making more and more cities construct high-rise and super high-rise buildings. Table 1 shows the number of super high-rise buildings in cities in China from 2013 to 2018. Because the load of the superstructure of high-rise and super-high-rise buildings is much larger than that of existing ordinary buildings, the high-rise and super-high-rise buildings have relatively high requirements for settlement deformation and bearing performance. At present, the pile foundation as a deep foundation form is being widely used in high-rise and super high-rise building foundations [1]. But a new problem arises: differential settlement of the foundation. The dish-shaped differential settlement problem caused by the traditional design method of piled raft (box) foundation of high-rise buildings and the differential settlement problem of the main podium [2]. "Technical Specifications for Building Pile Foundations" [3] proposed a new concept of variable stiffness leveling design. The basic idea of this concept is to consider the joint action of the foundation, foundation and superstructure, and adjust the distribution of the supporting stiffness of the pile and soil, the dominant factor that affects the settlement and deformation field.

Table 1. Statistics of the number of super high-rise buildings in each city.

Ranking	City	Quantity	Ranking	City	Quantity
1	Tianjin	20	6	Wuhan	7
2	Shenzhen	15	6	Shenyang	7
3	Wuxi	9	6	Guiyang	7
4	Guangzhou	8	10	Chongqing	6
4	Dalian	8	10	Kunming	6
6	Shanghai	7			

Variable stiffness leveling is to adjust the stiffness distribution of the foundation and base, so that the load of the superstructure and the reaction force of the foundation are coordinated with each other, and the settlement of the foundation will tend to be uniform, so that the internal forces of the superstructure and the foundation are kept within a reasonable range. There are many methods for leveling the stiffness of a piled raft foundation, such as variable pile length, variable pile diameter, variable pile spacing, combination of variable pile length and variable pile diameter, combination of variable pile length and variable pile distance, variable pile diameter and variable The combination of pile spacing and the combination of variable pile length, variable pile diameter, and variable pile spacing. Can be expressed by a general equilibrium equation [4]:

$$\left[K_b + K_r + K_s\right]\left\{\mu_0\right\} = \left\{Q\right\} + \left\{S_b\right\} \tag{1}$$

Where： $[K_b]$ is equivalent boundary stiffness matrix of superstructure; $[K_r]$ is base stiffness matrix; $[K_s]$ is foundation stiffness matrix;$\{\mu_0\}$is displacement vector of base node;$\{Q\}$is load vector of base part node;$\{S_b\}$ is equivalent node load vector of superstructure.

It can be seen from equation (1) that the settlement of the foundation is directly related to the stiffness of the superstructure, foundation, and foundation. To reduce the differential settlement of the foundation, it is necessary to increase the stiffness of the above three. Among them, the effect of increasing the stiffness of the superstructure on reducing the differential settlement of the foundation is not obvious, and the effect of increasing the stiffness of the foundation on reducing the differential settlement of the foundation is also not good, and it will also increase the cost. Therefore, only by adjusting the foundation stiffness K_r can the differential settlement of the foundation be effectively reduced.

In the research field of variable stiffness leveling of piled raft foundations, Liu Jinli [5] proposed the concept and method of optimal design for joint variable stiffness leveling to adjust the pile length on the basis of maintaining the same amount of raw materials to reduce the differential settlement of the foundation. . Franke et al. [6] broke through the traditional design method and used unequal pile-length combed pile foundations (internally strong and weakly outside the pile) to reduce uneven settlement and prevent the building from tilting, and achieved good results. Basuony et al. [7] used model tests to analyze the bearing and settlement properties of single piles, raft foundations, and piled raft foundations in sandy soils. At the same time, the slenderness ratio of the piles, the relative stiffness of the rafts, and the rafts were analyzed. Effect of thickness on load bearing and settlement properties. Ghalesari et al. [8] designed an optimal pile length and pile position layout scheme through specific examples, and pointed out that the economic and practical design of the pile raft foundation was carried out by optimizing the pile length and pile position. Abdrabbo et al. [9] studied the effects of piles with different pile lengths and diameters on the working performance of group piles through model tests.

In order to explore the effects of different combination methods on differential settlement and internal force, this article studies the superstructure and foundation as a whole, and analyzes the variable pile length, variable pile diameter, combination of variable pile length and variable pile diameter, and centrally-distributed piles.

2 ABAQUS finite element analysis method and model of pile foundation

With the continuous development of computer technology and the emergence and promotion of commercial finite element software, numerical simulation analysis methods are becoming more and more popular with scholars. For the study of pile-soil interaction, you only need to set relevant parameters and establish corresponding models according to the research needs, and you can calculate the results that are more realistic and basically consistent with the field test, which greatly reduces the time and cost of scientific research. To achieve a more satisfactory result.

ABAQUS can be used to analyze a variety of mechanical model problems, whether linear or more complex non-linear problems. ABAQUS can be divided into ABAQUS / Standard (implicit method) and ABAQUS / Explicit (explicit method) two modules, of which ABAQUS / Standard is a standard general analysis module, which can be used to analyze static, dynamic, and stress-coupled analysis problems. ABAQUS / Explicit is mainly used to simulate and analyze relatively short-term transient dynamic problems, such as earthquakes, explosions, and shocks. The study of pile-soil interaction in group pile foundations in this paper mainly involves static problems, that is, to simulate the working state of group pile foundations under static load, so the ABAQUS / Standard module can be used. Numerical simulation of pile-soil interaction can be completed through ABAQUS / CAE (pretreatment) and ABAQUS / VIEWER (posttreatment).

2.1 Introduction to computational models

The model of this example is: 12-story frame-core tube superstructure with a height of 3m, the core tube shear wall thickness is 250mm, the beam section size is 250mm × 600mm, and the column section size is 600mm × 600mm. The raft is 0.6m thick and its plane size is 26.4m × 26.4m. The pile is a reinforced concrete pile with a diameter of 600mm and a length of 19m. Concrete material parameters in the model: Poisson's ratio is 0.2, density is 2500Kg/m^3, and elastic modulus is 30GPa. The layout of the pile is shown in figure 1.

Modeling with ABAQUS finite element software, taking a range of 132m × 132m × 53m of ground soil. The constitutive model of the foundation soil is Mohr-Coulomb's elastoplastic model. At the same time, a horizontal acceleration of 0.05g was applied to the superstructure to simulate earthquake loads. See table 2 for material parameters.

Table 2. Calculation parameters of soil.

Name of the soil	Thickness (m)	Elastic modulus E(MPa)	Poison's ratio μ	Cohesion C(kPa)	Friction angleΦ(°)	densityρ (Kg/m^3)
Silty clay	53	31.5	0.3	30.0	20.0	1800

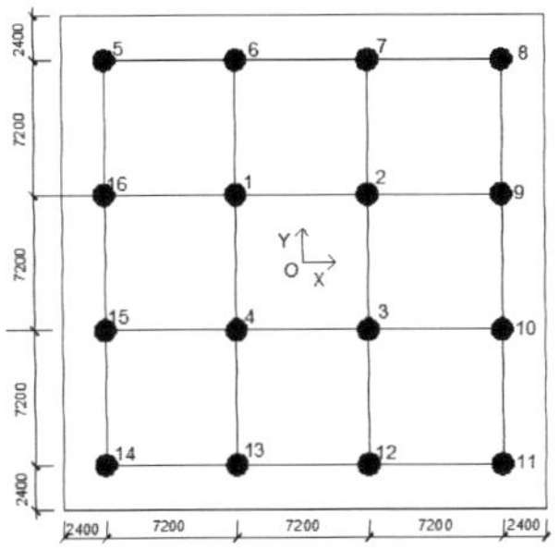

Figure 1. Layout plan of the piles.

2.2 Model mesh and boundary conditions

In this model, beam elements are used to simulate the beams and columns in the superstructure, shell elements are used to simulate the shear walls of the floor slab and the core tube, and the pile raft foundation and soil are simulated by solid elements. Considering the interaction, the superstructure is rigidly connected to the raft, and the coefficient of friction between pile and soil is 0.3. The side of the soil is restrained laterally, and it can slide freely vertically. The bottom surface is completely restrained in three directions. As shown in figures 2 and 3.

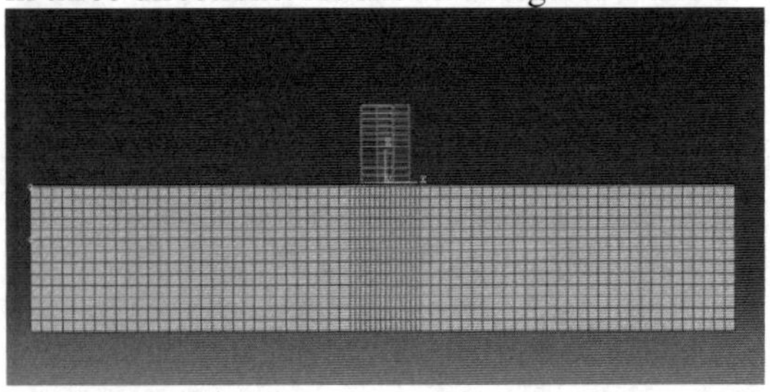

Figure 2. Holistic model.

Figure 3. Core tube.

3 Study on the influence of different forms of piles

3.1 Research on the influence of different pile length on differential settlement

Change the pile length of No. 1 ~ 4 pile to 29m, and other parameters are not changed, that is, change the pile length to adjust the foundation stiffness distribution. Through finite element calculation, the OX axis settlement distribution of the centerline of the raft is shown in figure 4 and the axial force distribution of No. 1 to 4 center piles is shown in table 3.

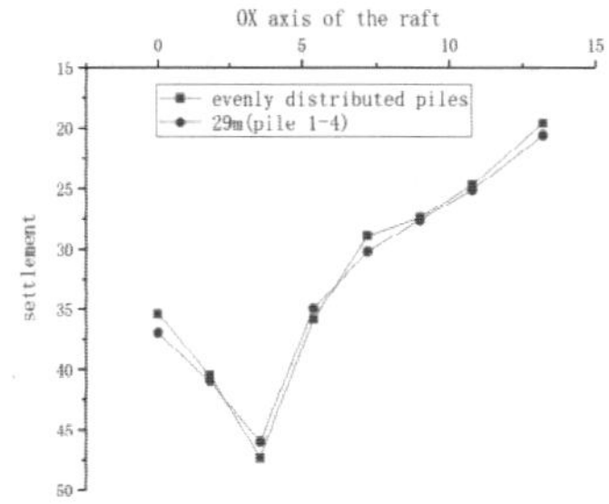

Figure 4. OX axis settlement map of the raft.

Table 3. Axial force of the center piles'top (KN).

Form of piles	NO.			
	1	2	3	4
Evenly distributed piles	1294	1362	1292	1219
Different pile's length	1440	1508	1453	1393

It can be seen from figure 4 that, compared with the uniformly-distributed piles, the variable pile length reduces the differential settlement of the foundation. When the piles are evenly distributed, the differential settlement of the raft is 22.6mm. When the length of the central pile is 29m, the differential settlement of the foundation is 14.6mm, which is 8mm less than that of the uniform pile. At the same time, it can be seen from table 3 that the axial force of the top of the center pile at the same location increases. This is because the stiffness of the foundation is adjusted by increasing the length of the center pile, and the internal force distribution of the foundation is also changed. Axial force at the top of the pile increases.

3.2 Research on the influence of different pile diameters on differential settlement

Take the diameter of the center pile (No. 1 ~ 4 piles) as 0.8m, and other parameters remain unchanged. Adjust the level of stiffness by increasing the diameter of the center pile. By finite element calculation, the OX axis settlement distribution of the centerline of the raft is shown in figure 5 and the axial force distribution of No. 1 to 4 center piles is shown in table 4.

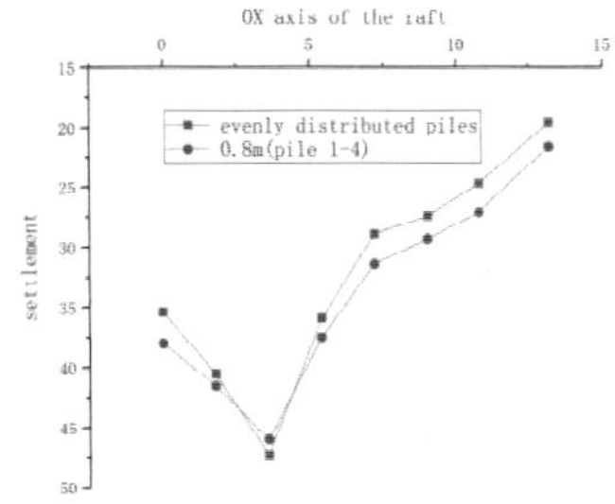

Figure 5. OX axis settlement map of the raft.

Table 4. Axial force of the center piles'top (KN).

Form of piles	NO.			
	1	2	3	4
Evenly distributed piles	1294	1362	1292	1219
Different pile's diameter	1534	1609	1545	1470

As can be seen from the figure, compared to uniformly-stacked piles, the differential settlement of the rafts is reduced by the form of piles with varying center-piled diameters, and the differentially-settlement of the uniformly-rafted rafts is 22.6mm. When the diameter of the central pile is 1.2m, the differential settlement of the raft is 20.1mm, which is 2.5mm less than that of a uniform pile. As can be seen from the table, the axial force of the top of the pile in the form of a variable center pile diameter is larger than the top axial force of the uniformly piled pile. This is due to the increase of the center pile diameter and the readjustment of the stiffness of the foundation. The internal force of the foundation is reduced, the raft settlement at the center pile is reduced, and the axial force at the top of the pile is increased.

3.3 Research on the influence of different pile spacing on differential settlement

By increasing the number of piles under the core tube, the pile spacing is reduced. Add four piles at the lower part of the core tube, as shown in figure 6, the remaining parameters remain unchanged. By finite element calculation, the OX axis settlement distribution of the centerline of the raft is shown in figure 7 and the axial force distribution of No. 1 to 4 center piles is shown in table 5.

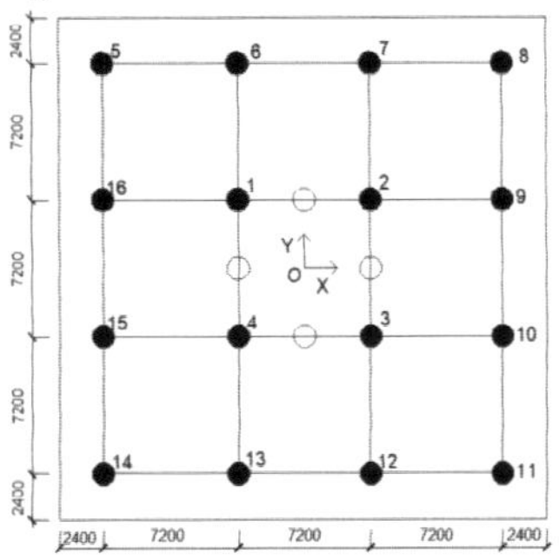

Figure 6. Layout plan of piles.

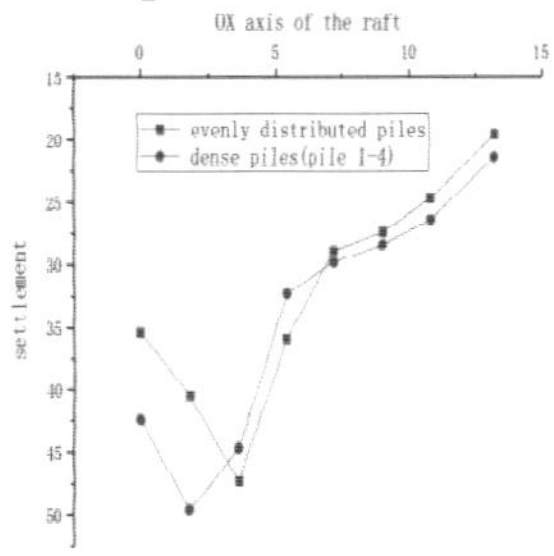

Figure 7. OX axis settlement map of the raft.

Table 5. Axial force of the center piles'top (KN).

Form of piles	NO. 1	2	3	4
Evenly distributed piles	1294	1362	1292	1219
Dense piles	1112	1235	1195	1039

From the results, it can be seen that compared with uniformly piled piles, the differential settlement of the piled piles in the form of reduced pile spacing is reduced, and the differential settlement of the uniformly piled raft is 22.6mm. The differential settlement of the pile foundation in the form of dense piles is 18.3mm, which is 4.3mm smaller than that of uniform piles. It can be seen from the table that the axial force of the central pile and the uniformly-distributed pile in the form of a dense pile are reduced. This is due to the addition of four piles in the center of the raft to share the upper load.

3.4 Research on the influence of different settlement of different pile length and pile diameter combinations

The central piles (No. 1 ~ 4 piles) are 24m long and 1.0m diameter. Adjust the stiffness of the foundation by changing the pile length and diameter. By finite element calculation, the OX axis settlement distribution of the centerline of the raft is shown in figure 8 and the axial force distribution of No. 1 to 4 center piles is shown in table 6.

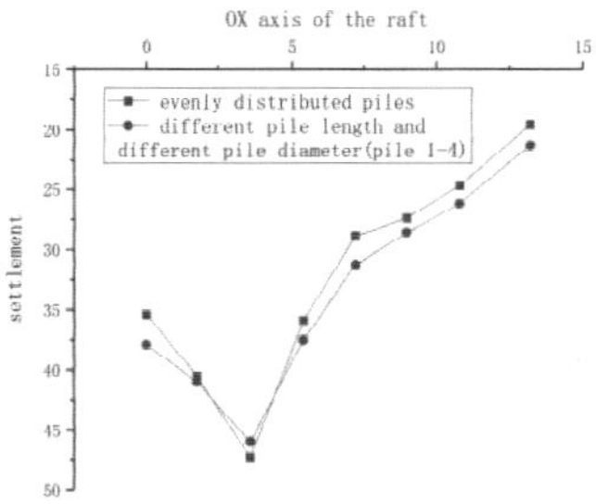

Figure 8. OX axis settlement map of the raft.

Table 6. Axial force of the center piles'top (KN).

Form of piles	NO. 1	2	3	4
Evenly distributed piles	1294	1362	1292	1219
Different pile lengths and different pile diameters	1477	1541	1483	1405

As can be seen from the figure, compared with the uniformly piled piles, the piled piles with different pile lengths and different pile diameters reduce the differential settlement of the raft. The differential settlement of the uniformly piled raft is 22.6mm. When the central pile length is 24m and the pile diameter is 1.0m, the differential settlement is 21mm, which is 2.5mm less than the uniform pile distribution. As can be seen from the table, the stiffness of the foundation is adjusted by the form of piles with different combinations of pile lengths and diameters. The internal force of the foundation also changes.

4 Conclusion

According to the calculation results, it can be found that the average settlement of the variable pile length, variable pile diameter, compacted pile, combined pile length and variable pile diameter in the form of pile piles is reduced compared to uniform pile piles, but the reduction is not significant enough. Although the above pile-laying forms do not have a good effect on reducing the average settlement, the differential settlement of pile-raft rafts with reduced pile length, variable pile diameter, dense pile, and combination of variable pile length and variable pile diameter has been reduced, which has achieved relatively good result.

In addition, among these several types of piles, the variable pile length has the best control effect on differential settlement, followed by the variable pile spacing type, variable pile diameter, and the combination of variable pile length, and variable pile diameter. The effect of the pile form is relatively poor. In terms of material usage, the material used for the variable pile length and variable diameter pile form is much less than the other two types of pile forms. Among them, the variable pile length form has a better effect on reducing differential settlement than the variable pile distance, and the amount of material And less. Therefore, under certain conditions, the material used in the form of variable pile length is more economical, and the optimal design effect of the foundation is also better. At the same time, in actual engineering, which form of piles to be used needs to be considered in combination with specific site conditions and construction difficulty, as well as economic applicability, etc.

5. References

[1] Cui Tao. Numerical Analysis and Settlement Calculation of Variable Stiffness Pile Group Foundation [D]. *Harbin Institute of Technology*, 2016.

[2] Wang Tao. Research on bearing behavior of soil under raft bottom under variable stiffness levelling pile mode [J]. *Chinese Journal of Geotechnical Engineering*,2011,**33(07)**:1014-1021.

[3] JGJ94—2008 Technical specifications of building pile foundation [S]. 2008. (JGJ94—2008 Technical code for building pile foundation[S]. 2008. (in Chinese).

[4] Liu Jinbo, Huang Qiang. Understanding and Application of Technical Specifications for Building Pile Foundations [M]. *China Construction Industry Press*, 2008.

[5] Liu Jinli, Chi Lingquan. Pile-soil deformation calculation model and variable stiffness leveling design [J]. *Chinese Journal of Geotechnical Engineering*, 2000**(02)**:151-157.

[6] Franke E, Lutz B, El-Mossallamy Y. Measurements and numerical modelling of high rise building foundations on Frankfurt Clay[C]. *Vertical and Horizontal Deformations of Foundations and Embankments*. ASCE, 1994: 1325-1336.

[7] Basuony E G, Ahmed A G, Abdel F Y, et al. Behavior of raft on settlement reducing piles: Experimental model study[J]. *Journal of Rock Mechanics and Geotechnical Engineering*, 2013, **5(5)**: 389-399.

[8] Ghalesari A T, Barari A, Amini P F, et al. Development of optimum design from static response of pile–raft interaction[J]. *Journal of Marine Science and Technology*, 2015:1-13.

[9] Abdrabbo F M, El-wakil A Z. Behavior of pile group incorporating dissimilar pile embedded into sand[J]. *Alexandria Engineering Journal*, 2015, **54(2)**: 175-182.

IOP Conf. Series: Materials Science and Engineering **829** (2020) 012026 doi:10.1088/1757-899X/829/1/012026

Research on Magnetic Induction Detection of Steel Bars Corrosion

Zengquan Yang[1], Jinsong Xiong[2], Chuan Chen[2], JunXu[2], Jiulin Zhang[2]and Huarui Ye[3*]

[1] Ningde shachengwan cross-sea Expressway Co., Ltd, Ningde, Fujian, China
[2] Chongqing Hongyan Construction Machinery Manufacturing Co., Ltd, Beibei, Chongqing, China
[3] State Key Laboratory of Power Transmission Equipment & System Security and New Technology, Chongqing University, Shapingba, Chongqing, China

*739148755@qq.com

Abstract. Reinforced concrete structures are widely used in modern buildings. However, the structural damage caused by the steel bars corrosion seriously affects the building remaining life. The reinforcement corrosion in concrete is slow, so the long-term detection of reinforcement corrosion is very important. The development of reinforcement corrosion detection equipment in China is slow, which cannot meet the application requirements of domestic concrete structures, this paper proposes a non-contact magnetic induction detection method for reinforcement corrosion. Through theoretical analysis and experimental design, a simple model is built and simulated. Simulation results of single and multiple steel bars corrosion to verify the feasibility of the method. This paper provides theoretical support and technical solutions for the operation state detection of non-contact reinforced concrete, and makes contribution to the non-contact magnetic induction detection technology.

1. Introduction

Reinforced concrete structure is an essential component of modern architecture. No matter in infrastructure or science industry, reinforced concrete structure is the top priority. However, due to the difference of application environment and people's low awareness of reinforced concrete protection, structural failure or even collapse accident will occur, which brings great threat to people's safety and causes economic losses. Professor Mehta from Berkeley University summed up several important factors in the concrete structures destruction in the world at the second International Conference on concrete durability. He pointed out that the primary factor affecting the durability of concrete structures was the steel bars corrosion [1]. Therefore, the detection and research of reinforcement corrosion has become an important problem in today's society [2].

At present, there are many detection and research methods for steel bars at home and abroad. According to the damage to the structure, there are destructive detection and non-destructive detection. According to the detection principle, there are analytical method, physical method and electrochemical method [3]. The analysis method is to comprehensively analyse various factors of the tested object environment, such as temperature, humidity and pH value, which will affect the corrosion rate of steel bars, and predict the steel bar corrosion by modelling. Compared with the destructive detection, the advantage is low cost and no waste, but the disadvantage is too many factors affecting the steel bars corrosion, which cannot be considered comprehensively. Besides, lacking of theoretical support and the low reliability are also affect the results. There are many physical detection methods, but the existing detection methods can only be used for qualitative analysis, mainly used in the laboratory testing stage, which existence large errors in the actual measurement. As for the electrochemical detection process, there must be signal injection. On the one hand, it will damage the reinforced concrete in a certain extent, on the other hand, the signal cannot be injected into the corroded reinforcement under some working conditions.

In view of the above problems, this paper proposes a non-contact magnetic induction detection method for corroded steel bars. Firstly, this paper explains the non-contact magnetic induction method, and introduces some principles of the non-contact magnetic induction method. Secondly, the basic principle of contactless magnetic induction experiment is described. Then, according to the internal reinforced concrete structure, a simple magnetic induction model is built by using the finite element simulation software COMSOL. Considering the actual effect of magnetic induction, the single corrosion and multiple corrosion simulation are carried out under the single-layer reinforced structure by injecting pulse excitation current. Comparing the magnetic field intensity distribution simulation on the coil and the reinforcement surface, combining the simulation results and the actual experiment of the reinforcement, under the appropriate excitation frequency, the reinforcement corrosion state can be effectively distinguished. It provides the theoretical support and technical solution for the operation state detection, and provides the foundation for the non-contact magnetic induction detection technology.

2. Non-contact magnetic induction measurement method and basic principle

In order to realize the high-speed and high-efficiency automatic detection of the steel bars corrosion degree without damaging the reinforced concrete. In this paper, a non-contact magnetic induction detection method is proposed. Its measurement method and basic principle are as follows.

2.1. Non-contact magnetic induction measurement method

The non-contact magnetic induction measurement method is shown in figure 1,.the universal non-contact reinforcement corrosion detection device can move freely on the concrete surface, generate eddy current effect on the reinforcement material in the concrete through the excitation coil, then detect the eddy current signal through the detection coil. Finally, it can judge the reinforcement material corrosion degree through experimental comparison and analysis [4, 5].

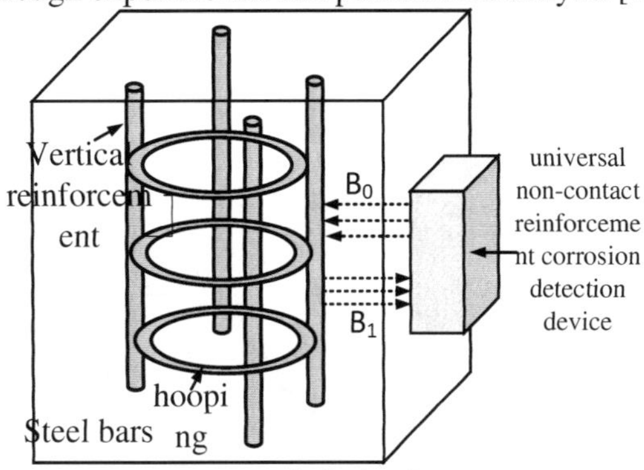

Figure1. Non-contact magnetic induction measurement method.

When the metal conductor cuts the magnetic line in the magnetic field or places in a changing magnetic field, according to Faraday's electromagnetic induction theorem, the metal conductor will feel a current similar to the water vortex, which is called eddy current because of its vortex shape. Compared with the traditional eddy current detection technology, the pulse eddy current detection technology loads the pulse current signal with a certain duty cycle and frequency on the excitation coil. When the excitation current signal is turned off instantaneously, the excitation coil induces a pulse magnetic field which decays rapidly with time, which called a primary field. The fast changing primary field induces a vortex like pulse current inside the conductor specimen. The pulse eddy current signal attenuates with time propagation in the conductor and induces a new changing magnetic field, which is called the secondary field, as shown in figure 2. According to the principle of electromagnetic induction, when the conductor material is detected by the detector probe, the detector coil output is the transient voltage induced by the decay of the magnetic field around the coil with time [6].

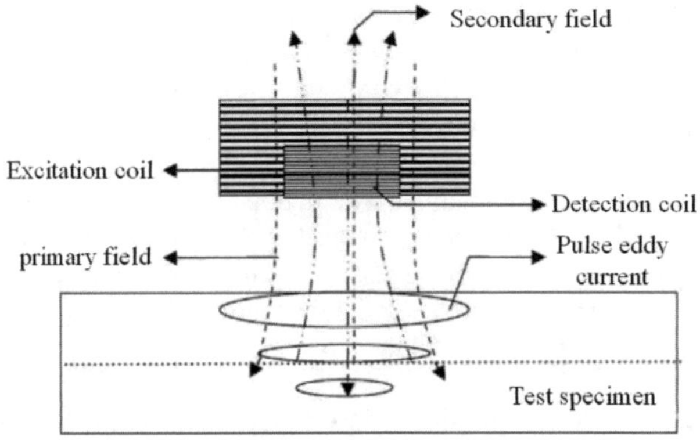

Figure 2. Basic principle of pulsed eddy current testing.

The pulsed eddy current testing technology use the eddy current effect physical phenomenon. When the tested steel bar is corroded and its thickness and outer surface structure are changed, the induced eddy current in the steel bars and the magnetic field distribution are bound to be affected by the corrosion and change accordingly. By analysing these changes, we can master the corrosion situation in the steel bar. Therefore, the information such as the conductivity, magnetic conductivity or thickness of the corroded steel bars will have a direct impact on the induction signal. Through the data collection, processing and analysis of the induction voltage signal, the corresponding corroded steel bars corrosion information can be mastered.

2.2. Basic principle of non-contact magnetic induction testing for corroded steel bars s

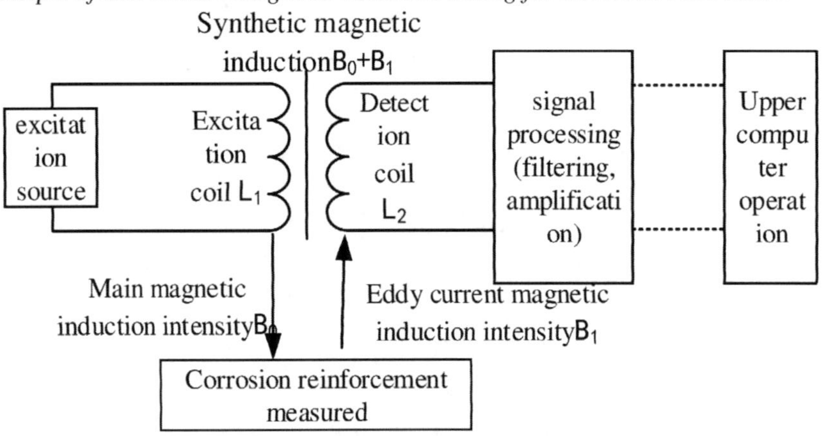

Figure 3. Basic principle of magnetic induction testing for non-contact corrosion reinforcement.

In this paper, the basic principle of non-contact corrosion reinforcement magnetic induction detection is shown in figure 3. The pulse excitation signals with different fundamental frequency, duty cycle and amplitude are generated by the excitation source. Then input the excitation signal into the excitation coil $L1$, the excitation coil will generate a magnetic field in space, and the reinforcement in reinforced concrete will generate eddy current effect under the space magnetic field [7]. Assuming that the medium in the field is linear isotropic, the eddy current electric field E can be expressed as follows:

$$\vec{E} = -j\omega\vec{A} - \nabla\varphi \tag{1}$$

Where, ω is the angular frequency, A is the magnetic vector potential, and φ is the scalar potential. The area where eddy current exists is passive, and the eddy current density meets the following formula:

$$\nabla\cdot\vec{J}=0 \tag{2}$$

The formula (2) is substituted into the differential form $\vec{J}=\sigma\cdot\vec{E}$ of Ohm's theorem, then the result is substituted into the formula and use the vector operation. Finally, the Coulomb specification $\nabla\cdot\vec{A}=0$ is used:

$$\nabla\cdot(\sigma\nabla\varphi)=-\omega\vec{A}\cdot\nabla\sigma \tag{3}$$

According to the condition that the normal component of current density is continuous on the interface between concrete and reinforcement, the boundary condition and the interface connection condition between concrete and reinforcement can be obtained:

$$\frac{\partial\varphi}{\partial n}=-\omega A_n \tag{4}$$

$$\sigma_2\frac{\partial\varphi_2}{\partial n}-\sigma_1\frac{\partial\varphi_1}{\partial n}=-\omega\left(\sigma_2 A_{2n}-\sigma_1 A_{1n}\right) \tag{5}$$

Where, $\vec{A}$ is the component of the magnetic vector position in the normal direction outside the boundary of the medium. The above three formulas constitute the partial differential equation and the definite solution condition of the vortex flow field in the field. The distribution of eddy current field can be obtained by (6):

$$\vec{J}=\sigma\cdot\vec{E}=\sigma\cdot(-j\omega\vec{A}-\nabla\varphi) \tag{6}$$

The magnetic induction intensity in the measurement coil consists of two parts: the magnetic induction intensity $\vec{B_P}$ generated by the current in the excitation coil: the magnetic induction intensity $\vec{B_S}$ generated by the eddy current in the reinforcement. That is $\vec{B}=\vec{B}_P+\vec{B}_S$. $\vec{B_P}$ basically unchanged, so the magnetic induction intensity change in the measuring coil is the change of $\vec{B_S}$. That is $\Delta\vec{B}=\Delta\vec{B}_S$. According to Biot-Savart Law:

$$\vec{B}_S=-j\frac{\mu_0}{4\pi}\int\vec{J}\times\frac{\vec{R}}{R^3}dV=-j\frac{\mu_0}{4\pi}\int\sigma(\omega\vec{A}+\nabla\varphi)\times\frac{\vec{R}}{R^3}dV \tag{7}$$

In the above formula, $\vec{R}$ is the vector from the source point to any field point in the reinforcement. In order to determine the corresponding relationship between the magnetic induction intensity change $\Delta\vec{B}_s$ and the conductivity change $\Delta\sigma$, it is assumed that $\vec{B_S}$ and $\vec{B_{S0}}$ are the magnetic induction intensities when the conductivity is σ and σ_0 , $\varphi_0+\Delta\varphi$, $\sigma_0+\Delta\sigma$ instead of the φ and σ in the above formula, the following formula is obtained:

$$\vec{B_S}=-j\frac{\mu_0}{4\pi}\int(\sigma_0+\Delta\sigma)\left[\omega\vec{A}+\nabla\left(\varphi_0+\Delta\varphi\right)\right]\times\frac{\vec{R}}{R^3}dV \tag{8}$$

Substituting the above formula into $\Delta\vec{B}_s=\vec{B}_s-\vec{B}_{s0}$, ignoring the last term of the above formula, and substituting $\frac{\partial(\nabla\varphi)}{\partial\sigma}\Delta\sigma$ with $\nabla(\Delta\varphi)$, we can get the following corresponding relationship with the change of $\Delta\vec{B}_s$ and conductivity $\Delta\sigma$:

$$\Delta\vec{B}_S=-j\frac{\mu_0}{4\pi}\int\left[\left(\omega\vec{A}+\nabla\varphi_0+\sigma_0\frac{\partial(\nabla\varphi)}{\partial\sigma}\right)\Delta\sigma\times\frac{\vec{R}}{R^3}\right]dV \tag{9}$$

The above formula explains the relationship between the reinforcement conductivity change and the magnetic field change in the measuring coil. In the actual measurement, the system records the induction electromotive force in the detection coil, so only the relationship between the induction electromotive force change and the conductivity change needs to be clarified. Through the detection of electromotive force change, the data is transmitted to the upper computer, and the upper computer background software realizes data recording and analysis, which can complete the judgment of the steel bars corrosion degree.

3. Simulation and verification of contactless magnetic induction

3.1. Simulation model building

Figure 4 is the model diagram of single-layer grid reinforcement. In the figure, the ring is the excitation coil. Under the ring, there are 10 reinforcements arranged in a crisscross single layer. The diameter of a single reinforcement is 16mm and the length is 100cm. The distance between each two reinforcements is 1.5cm. The distance between the coil and the centre of the reinforcement is 12cm. The rectangle outside the reinforcement is concrete, and the largest cylinder outside is the boundary.

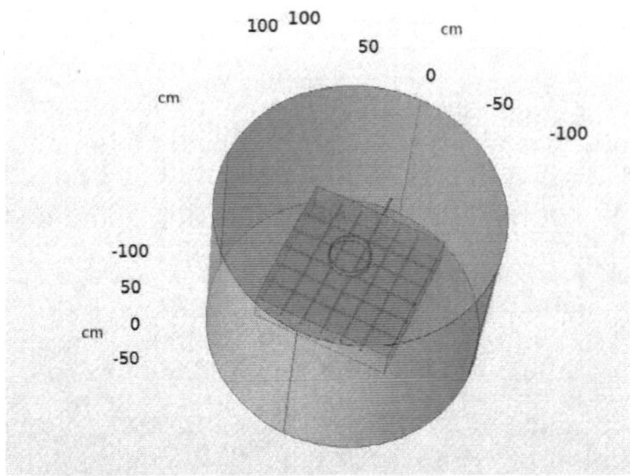

Figure 4. Grid reinforcement model.

3.2. Simulation of single steel bar corrosion under pulse excitation

At this time, it is simulated that one reinforcement in a single layer is corroded and the other reinforcement is in good condition. The number of coil turns is 1000, and the excitation is the pulse excitation signal with peak value of 1A. At this time, the radius of the steel bar is 10-16mm, and a parametric scan is made every 2mm. The change of the magnetic flux density with time at the point directly below the coil centre is shown in figure 5.

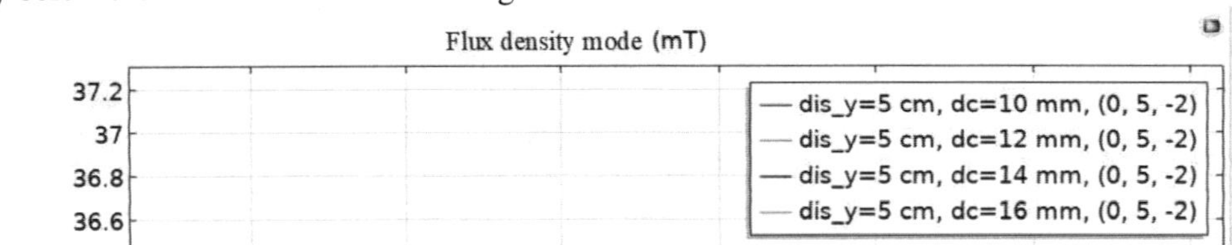

Figure 5. Change of flux density mode with time for different thickness of steel bars.

It can be seen from figure 5 that the corresponding magnetic flux density of different diameter steel bars increases first and then decreases with the pulse signal. The thicker the steel bars are, the maximum magnetic flux density increases with the increase of the steel bars radius. When the coil is directly above the steel bars, the steel bars corrosion degree can be determined by measuring its magnetic flux density mode. The coil power factor is shown in figure 6.

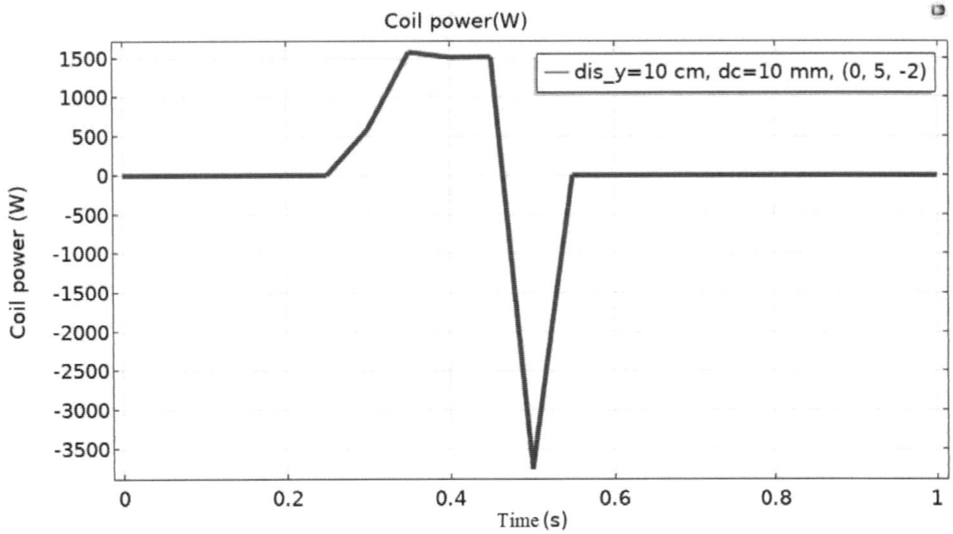

Figure 6. Change of coil power with time.

3.3. Simulation of coil position change relative to reinforcement under pulse excitation (single reinforcement corrosion)

As shown in figure 7, the positions of the coil positive center relative to the reinforcement are 10cm and 5cm (x=-10, -5) to the left, 10cm and 5cm (x=10,5) to the right and above (x=0) of one reinforcement respectively.

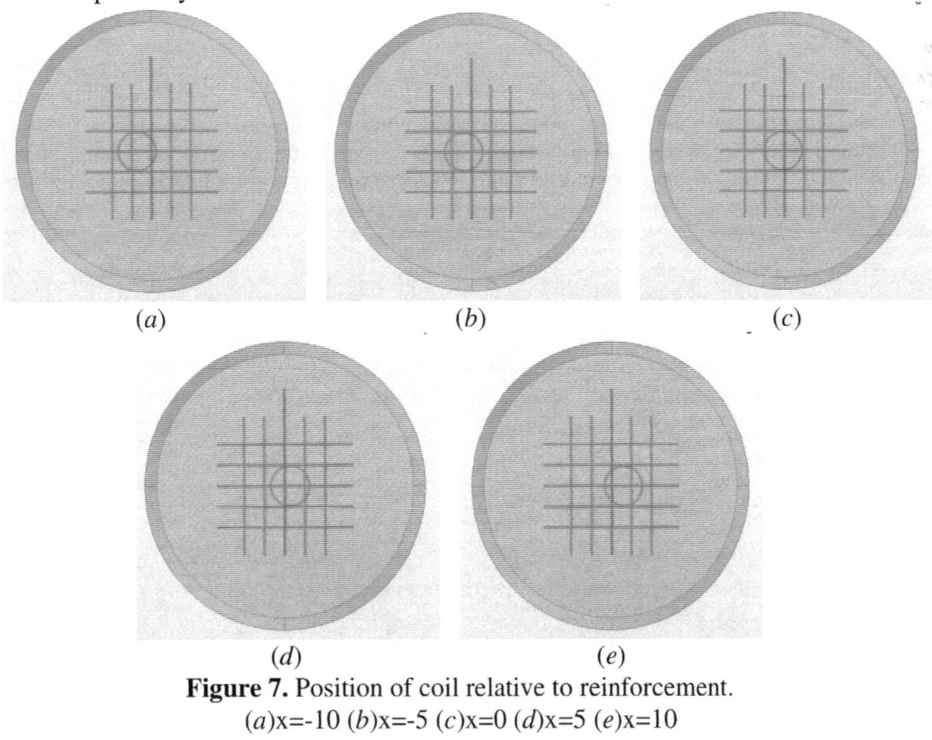

(*a*) (*b*) (*c*)

(*d*) (*e*)

Figure 7. Position of coil relative to reinforcement.
(*a*)x=-10 (*b*)x=-5 (*c*)x=0 (*d*)x=5 (*e*)x=10

IOP Conf. Series: Materials Science and Engineering **829** (2020) 012026 doi:10.1088/1757-899X/829/1/012026

Through simulation, the relationship between the coil position relative to the reinforcement and its corresponding flux density is analysed, and the results are in table 1.

Table 1. Peak flux density in different position.

Position of coil relative to reinforcement (cm)	Peak flux density (mT)
-10	36.35063491274944
-5	36.3139386890932
0	36.44302044251746
+5	36.29281354084986
+10	36.37778537627404

It can be seen from the figure that for the single corrosion steel bar, when the coil is directly above the steel bar (x = 0), its relative magnetic flux density is the largest, about 36.4mT, so when the coil moves slowly, the magnetic flux density mode reaches the maximum value, it can be judged that the steel bar is directly below the coil.
Through the simulation analysis of the single steel bar corrosion degree, it is found that the coil magnetic flux density is different at different positions above the steel bar and when the steel bar in different diameter. Through the effective analysis of the waveform, the coil position relative to the steel bar and the change of the steel bar thickness can be determined. That is to say, the steel bar corrosion degree can be determined by the magnetic flux density measurement

3.4. Simulation of multiple steel bars corrosion under pulse excitation

In the process of actual reinforcement corrosion, one piece of reinforcement will be corroded when it is corroded. Assuming there are four reinforcement corroded at the same time, and they are all under the coil. As shown in figure 8, four longer reinforcements are corroded.

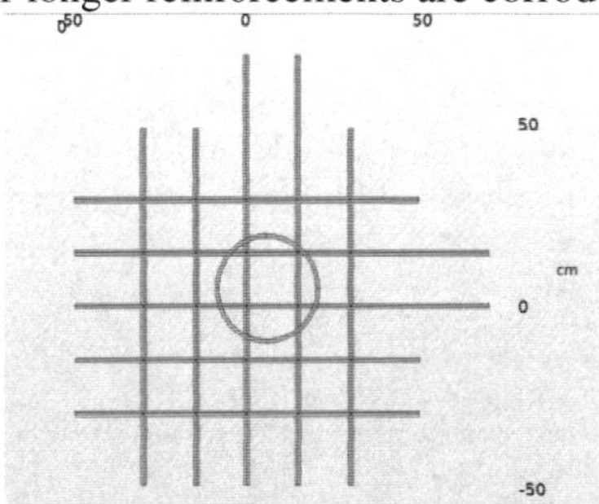

Figure 8. Uniform Corrosion of reinforcement.

At this time, the diameter of the steel bar is 10-20 mm, and a parametric scan is made every 2 mm. The change of the magnetic flux density in the centre of the coil with time is shown in figure 9.

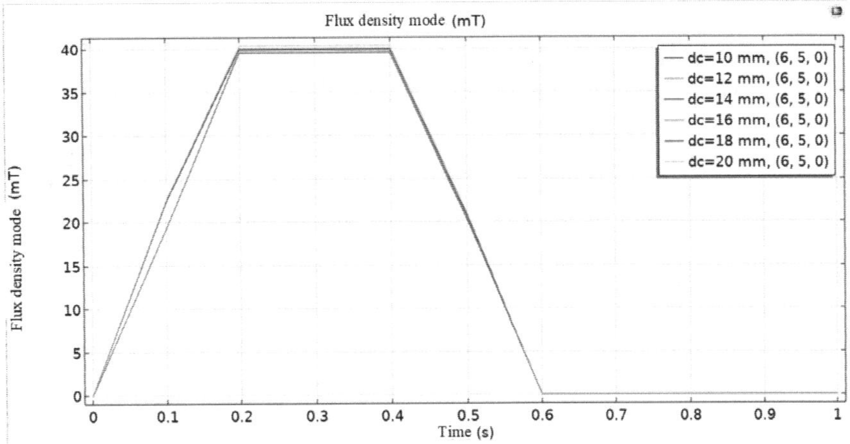

Figure 9. The flux density mode change with time for different diameter of uniform corrosion steel bars.

As shown in figure 9, the results of steel bars uniform corrosion are similar to the signals detected by a single. The detection signals corresponding to different diameter of steel bars are different. The corrosion degree of steel bars can be determined by detecting the magnetic flux density.

3.5. Simulation of the position change of the coil relative to the reinforcement under pulse excitation (multiple reinforcement corrosion)

As shown in figure 10, the positions of the coil positive centre relative to the reinforcement are 10cm and 5cm (x = - 10, - 5) to the left, 10cm and 5cm (x = 10, 5) to the right, and above (x = 0).

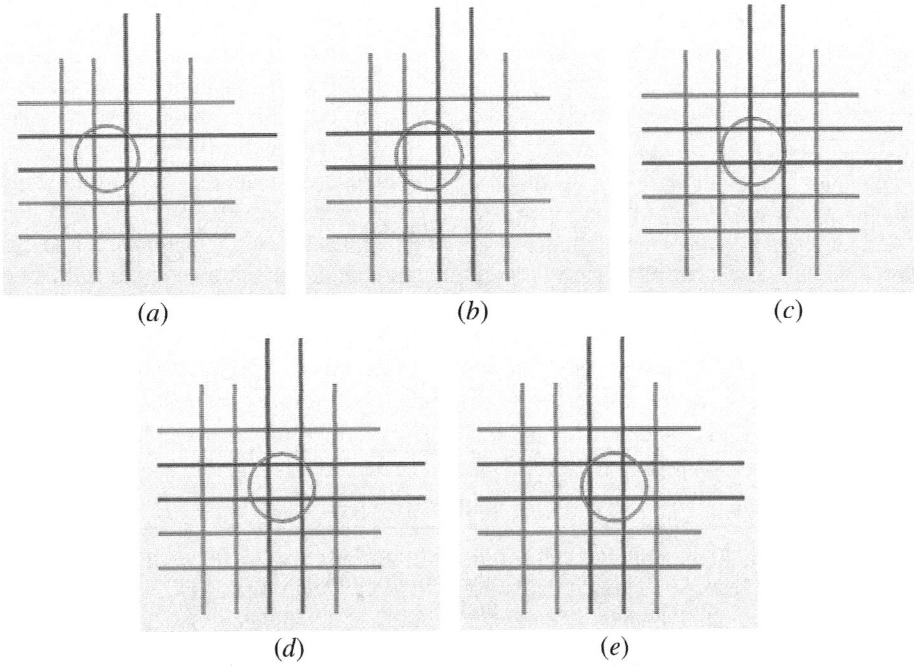

Figure 10. Position of coil relative to reinforcement.
(*a*)x=-10 (*b*)x=-5 (*c*)x=0 (*d*)x=5 (*e*)x=10

Through simulation to analyse the relationship between the position of the coil relative reinforcement and its corresponding magnetic flux density, the results are shown in table 2.

Table 2. Peak flux density in different position.

Position of coil relative to reinforcement (cm)	Peak flux density (mT)
-10	35.99007598205135
-5	35.957122385909436
0	36.101814549196774
+5	36.034059815755526
+10	36.05307547262015

It can be seen from table 2 that in the case of multiple steel bars corrosion, when the coil is directly above the steel bar (x = 0), its relative magnetic flux density is the largest, about 36.1mT, so when the magnetic flux density mode reaches the maximum value, it can be judged that the steel bar is directly below the coil at this time.

The single-layer and multi mesh reinforced structure will produce a lot of eddy current effects under the magnetic field. The superposition of multi magnetic fields will cause the signal waveform under the coil to be unstable. However, through the analysis of the relative magnetic flux density peak value, the reinforcement position and the corresponding corrosion degree can be determined.

4. Conclusion

Reinforced concrete material is a very important part of modern buildings, such as bridges, tunnels, various buildings and so on. In order to ensure the safety of the building, find out the possible problems in time, and make corresponding remedial measures to protect the national economy and people's life. The concrete structure quality detection technology has become a very important research topic in today's society. In this paper, the research status of concrete structure detection technology at home and abroad is introduced. By comparing the steel corrosion detection technology at present, the advantages of non-contact magnetic induction detection of corroded steel bars are obtained. Then, the non-contact magnetic induction method is introduced, including the pulse eddy current effect, excitation coil and detection coil electromagnetic parameter analysis theory. The basic principle and experimental scheme of non-contact magnetic induction detection method are studied. On the basis of a large number of principle analysis, a simple corrosion model of reinforced concrete is built for the reinforced concrete corrosion detection, and the simulation is carried out through the finite element simulation software COMSOL. It is concluded that the coil magnetic flux density is different at different positions above the steel bar and when the steel bar is of different diameter. The position of the coil relative to the steel bar and the change of the steel bar diameter can be judged by the waveform effective analysis, that is to say, the steel bar corrosion degree can be determined by the magnetic flux density measurement.

5. References

[1] Jha, Somesh, Chalasani, Prasad, Shehory, Onn et al. 1998 ACM Press the second international conference - Minneapolis, Minnesota, United States *J. Proceedings of the second international conference on Autonomous agents*, L457-458.

[2] Kenji Reichling, M. Raupach, J. Broomfield et al. 2013 Full surface inspection methods regarding reinforcement corrosion of concrete structures *GERMANY J. Materials & Corrosion*, **64**(2) L116-127

[3] CHUN Qing, PAN Jian-wu. 2014 Research on the methods for calculation and prediction of the service life of reinforced concrete buildings built during the period of the Republic of China in Jiangsu and Zhejiang provinces *China J. Sciences of Conservation & Archaeology*, **78**(1) L1436-1446.

[4] Tsukada, Keiji. 2018 Small Eddy Current Testing Sensor Probe Using a Tunneling Magnetoresistance Sensor to Detect Cracks in Steel Structures *United States J. IEEE Transactions on Magnetics* ,**54**(11)

[5] Tsukamoto, Akira. 2018 Development of Eddy Current Testing System Using HTS-SQUID on a Hand Cart for Detection of Fatigue Cracks of Steel Plate Used in Expressways *United States J. IEEE Transactions on Applied Superconductivity*, **28**(4) L 1-5.

[6] Fukuoka, Katsuhiro. 2019 Flaw detection for microcrack in spring steel and estimation of crack shape with eddy current testing *NETHERLANDS J. International Journal of Applied Electromagnetics and Mechanics*, **59**(4) L1267-1274.

[7] Kosukegawa, Hiroyuki. 2019 Characterization of laminated structure on scarfed slope of CFRP by utilizing eddy current testing with differential type probe *NETHERLANDS J. International Journal of Applied Electromagnetics and Mechanics*, **59**(4) L1227-1238.

IOP Conf. Series: Materials Science and Engineering **829** (2020) 012027 doi:10.1088/1757-899X/829/1/012027

Automatic Crack Detection for Concrete Infrastructures Using Image Processing and Deep Learning

Cuong Nguyen Kim[1 a], Kei Kawamura[2b], Hideaki Nakamura[3c], and Amir Tarighat[4d]

[1]Faculty of highway& bridge, Mien Trung of civil engineering, Vietnam
[2]Graduate School of Science & Technology for Innovation, Yamaguchi University, Japan
[3]Graduate School of Science & Technology for Innovation, Yamaguchi University, Japan
[4]Dept. of Civil Engineering, Shahid Rajaee Teacher Training University, Iran

[a] nguyenkimcuong@muce.edu.vn,[b]kay@yamaguchi-u.ac.jp, [c]Nakahide@yamaguchi-u.ac.jp, [d]tarighat@srttu.edu

Abstract. Automatic crack detection is a main task in a crack map generation of the existing concrete infrastructure inspection. This paper presents an automatic crack detection and classification method based on genetic algorithm (GA) to optimize the parameters of image processing techniques (IPTs). The crack detection results of concrete infrastructure surface images under various complex photometric conditions still remain noise pixels. Next, a deep convolution neural network (CNN) method is applied to classify crack candidates and non-crack candidates automatically. Moreover, the proposed method is compared with the state-of-the-art methods for crack detection. The experimental results validate the reasonable accuracy in practical application.

1. Introduction

Many concrete components of existing infrastructure systems such as bridges, and tunnels have suffered from various geologic, loading and environmental conditions cause to cracks which make influent to quality of operations. Therefore, the condition assessment of the existing infrastructures is an important task not only for warming against deterioration but also for guaranteeing soon maintenance. Concrete cracks are important indicators reflecting the safety of infrastructure. The automatic crack detection based on image data has been considered significantly due to accuracy, objectivity and timing inspection. This technique can be implemented using some of different image data captured from ultrasonic device, infrared and thermal device, laser scanning, and commonly digital cameras.

Image processing techniques consist of three approaches: edge detection, threshold technique (Fujita et al.2006) and mathematical morphology (Nguyen et al 2016). Machine learning algorithms (MLAs) are commonly used to decide the parameter value of IPTs. Therefore, they are applied to detect and classify concrete infrastructure surface cracks. In recently years, many automatic crack detection and classification methods based on a combination of IPTs and MLAs are implemented as decision tree (DT) (Kei et al.2013), support vector machine (SVM), k-clustering nearest neighbour (K-NN), and artificial neural network (ANN) (Zhang et al.2014).

Moreover, the parameter-optimization algorithms of IPTs such as genetic algorithm (GA), particle swarm optimization (PSO), artificial bee colony (ABC), and differential evolution (DE) are typically utilized.

In this article, the image processing parameters (IPPs) are adjusted to the optimized value in order to increase the accuracy of crack detection before use of convolution neural network (CNN) to eliminate non-crack candidates. As a result, the crack map will be improved significantly.

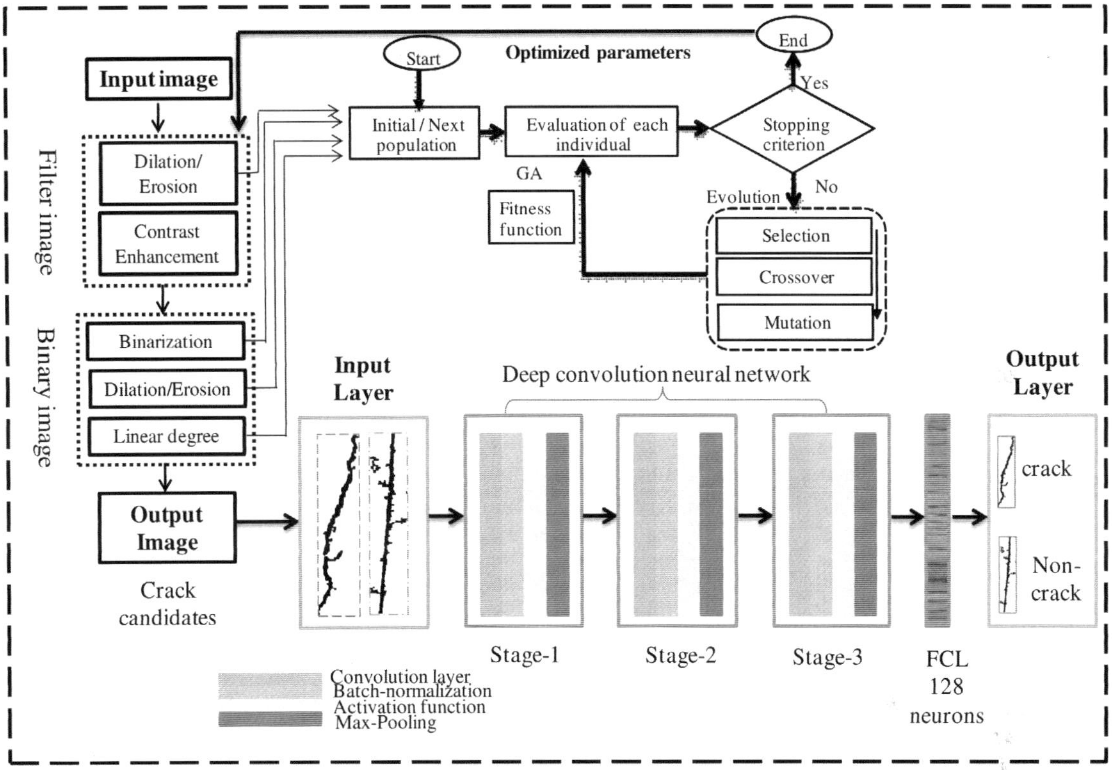

Figure 1. Pipeline of the automatic crack detection and classification using GA-CNN.

2. Proposed method

Figure 1 shows a pipeline of automatic crack detection and classification based on IPTs combined with image processing and deep learning.

2.1 Image Processing Techniques (IPTs)

The image processing techniques compose of three main parts. Namely, there are filtering image part, binary image part, and feature extraction part. Therein, filter image part comprises of morphological transform and contrast enhancement of the gray-scale image. The binary image part consists of binarization, and dilation-erosion transform of the binary image. To the end, geometric transform composes of labelling and linear degree.

2.1.1 Morphological transform. The morphology filter consists of opening transform of closing transform on gray-scale image with a predefined structuring element as the following equation:

$$M = \max([(G \oplus S)\ominus S]\ominus S]\oplus S], G) - G \qquad (1)$$

Where M means the smooth image after morphological image processing, S is a structuring element, G is a gray-scale image converting from the corresponded original image. Additionally opening transform includes dilation of erosion. Closing is an inverse operator of the opening as the following Eqs. (2) and (3):

$$(G \oplus S)_{(x,y)} = \max_{u,v}(G_{(x-u,y-v)} + S_{(u,v)}) \quad (2)$$

$$(G \ominus S)_{(x,y)} = \min_{u,v}(G_{(x-u,y-v)} - S_{(u,v)}) \quad (3)$$

The purpose of this part is firstly to smooth image as well as eliminate noises and shading and then make the sharpness of crack edge pixels. For an example, figure 2 (b) shows the result of smooth image *M*.

2.1.2. Structuring element design. Because the shape of crack is irregular elongation, the shape of the structuring element of line-type is adopted. To retrieve fully crack information in the various directions, combination line types of four directions into the structuring element such as {0°, 45°, 90°, 135°} or six directions {0°, 30°, 60°, 90°, 120°, 150°}. In this study, the former type is used to reduce computational volume.

2.2. Contrast enhancement

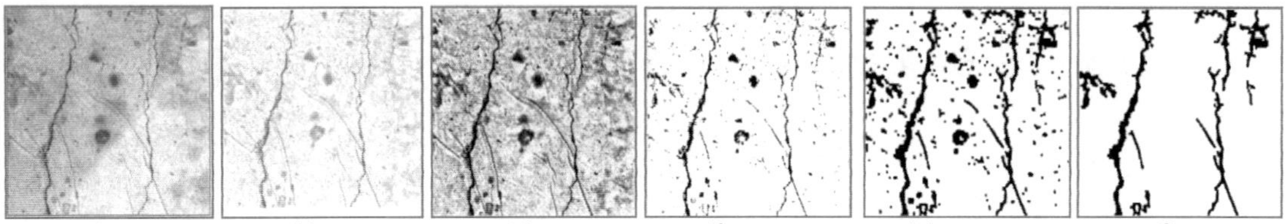

Figure 2. The illustrative results of image processing procedure. (a): Original image; (b): Morphological filter; (c): Contrast enhancement; (d):Binarization; (e): Dilation-thinning transform; (f): Final crack image processing.

Surface images of civil infrastructure are the typical dark environmental condition, narrow contrast. The intensity of crack pixels and background pixels is not much difference. Therefore, a histogram of the raw image is equalized before binarization to sharpen crack pixel as the following equation:

$$q_k = \sum_{i=0}^{k} \frac{n_i}{n} \quad k = 0,1,2,..L-1$$

$$I_k(out) = (I_{max} - I_{min}) \times q_k(in) + I_{min} \quad (4)$$

Where, n_i: is number of pixels that have i^{th} gray-scale value. n: the number of pixels; k: the input gray-scale level. L: the maximum gray-scale level (255). q_k: acquired normalized histogram. I_{max} and I_{min} are respective to the maximum and minimum intensities of the original image. $I_k(out)$ is the new gray-scale intensity of the input pixel having $q_k(in)$. As a result, figure 2.(c) shows crack enhancement after morphology filter is applied to smooth the original image.

2.2.1. Binarization. As shown in Eq. (5), the purpose of binarization is to segment gray-scale image into binary image. The binary image only has two values 255 (white pixel is background) and 0 (black pixel is crack or noise).

$$P(i) = \begin{cases} 0 \ \ if \ \ I(x_i) < T \\ 255 \ \ otherwise \end{cases} \quad (5)$$

Where T is a threshold value of the binarization (binary), it is necessary to find out an optimal value. P(i) is the ith pixel value after binarization step. figure 2(d) shows the results of binarization transform.

2.2.2. Dilation. The aim of a dilation operation is to connect crack fragments meanwhile noises are separated from the cracks. Therefore it results in the reduction of loss pixel. However, widths of crack pixels increase along with the crack shape. The size of structuring element of the dilation operation is considered as an adjusted parameter to optimized value by GA. In this paper, the shape of structuring element is predefined in "square" type. The rule of the dilation operation is if any pixel in the input neighborhood is "1", the output pixel is "1". Otherwise, the output pixel is "0".

2.2.3. Thinning. Thinning to a binary image is a morphology operator used to remove noise surrounding crack pixels. The purpose of this step is to prune branches from the crack shape as well as reduce noises after dilation operator is performed. Figure 2 (e) shows the image processing result of dilation and thinning transform.

2.2.4. Labeling and length threshold. After thinning, the pixel values of the image are presented by the 0s or 255s. The automatic labeling step is to combine the 8th-connection components as a single object with the same number. Hence, a single object can be a crack object or a noise object as shown in figure 2(f). Therefore, the length threshold decides as if the object is removed following equation:

$$P'^{(i)} = \begin{cases} 255\dfrac{Max(R_{xi}, R_{yi})^2}{S_i} < T' \\ 0 \ \text{otherwise} \end{cases} \quad (6)$$

Where T' is a threshold value of the ith label length to distinguish between a crack object and a noise object. S_i is the total number of pixels of the ith label. R_{xi}, and R_{yi} are the number of pixels of the ith label in the horizontal and vertical directions, respectively.

The purpose of the first part is to make blurred images as well as eliminate noises and shading. The main aim of the binarization is to segment grayscale image into binary image depending on a threshold value. The purpose of dilation/erosion that is processed in the binary image part is to connect fragment images of crack meanwhile noises are separated from the cracks. In the end, the threshold defined in linear degree is to decide whether the single objects in the binary image are removed.

The productivity of the filtered image depends on the parameter value of structuring element size in dilation and erosion transform. In binarization, if the threshold value is too high, many crack pixels are lost. If threshold value is too low, more noise will occur. Such it is necessary to find out an optimum threshold value. Similarly, the parameter values of dilation/erosion and linear degree in the binary image part also affect to the quality of output image. Therefore, these parameters are adjusted to the optimized values based on GA.

2.3. Application of GA to the image processing parameters optimization

2.3.1 Represented chromosome Design for solution candidates. The IPPs are combined together for creating an individual in a population. Next each individual is represented by a chromosome encoded to a binary string, as shown in figure 3.

Namely, the size of structuring element (s) is assigned by 6 bits, the threshold value of binarization (t) is expressed by 8 bits, dilation transform parameter (d) is expressed by 4 bits, and the linear degree is expressed by 6 bits (l). Table 1 shows the parameter value range which design based on the preliminary experiments.

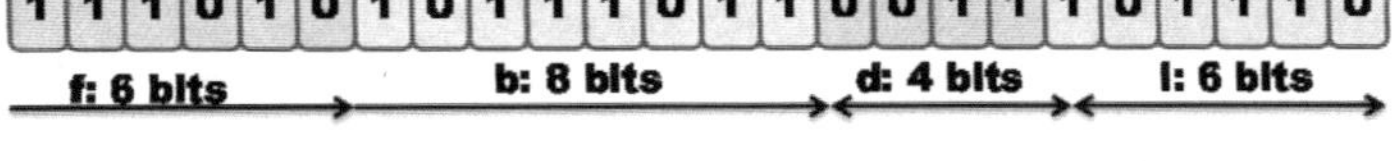

Figure 3. A represented chromosome for solution candidates.

IOP Conf. Series: Materials Science and Engineering **829** (2020) 012027 doi:10.1088/1757-899X/829/1/012027

Table 1. Properties of parameter.

Variable	Range	Steps	Bits
s	[1 127]	2	6
t	[0 255]	1	8
d	[1 31]	0.5	4
l	[0 32.5]	0.5	6

2.3.2. Genetic algorithm. Figure 4 indicates a sequence of GA including into the crucial three stages. Namely, they consist of the initial population generation, fitness evaluation of each individual in the current population, and evolution operation to create the next generation. Namely, the detailed steps are presented as the following three steps:

Step1. Generate initial population randomly

An initial population including 20 individuals was generated randomly with respect to 20 phenotypes to start fitness evaluation. To assess the fitness of the individual in the current population, an objective function to assess crack detection accuracy is defined as the Eq.(7). Loss and noise are computed based on comparison between the processed image and the target image shown in figure 5. As a result, the objective function (f) has to ensure the accuracy of extracted crack information with the minimum noises and losses as much as possible. The accuracy and the processing time can be considered as evaluation costs.

Step 2. Evolution operation

The evolution operation comprised of selection, crossover, and mutation is repeated until finding best solution. Each binary string encoded from the searching range of the parameter values has a corresponded fitness value. The probability of each string to be selected is proportional to its fitness value based on the Roultte wheel rotation randomly. The process is repeated for the second parent. Two elite members are kept forward to the next generation.

To improve quality of individual fitness, the crossover operation is used to create two new children from two selected parents with predefined probability. Crossover point is point laid on between 0 to the end of chromosome length. In this study, the single crossover point is selected. The part of the first parent

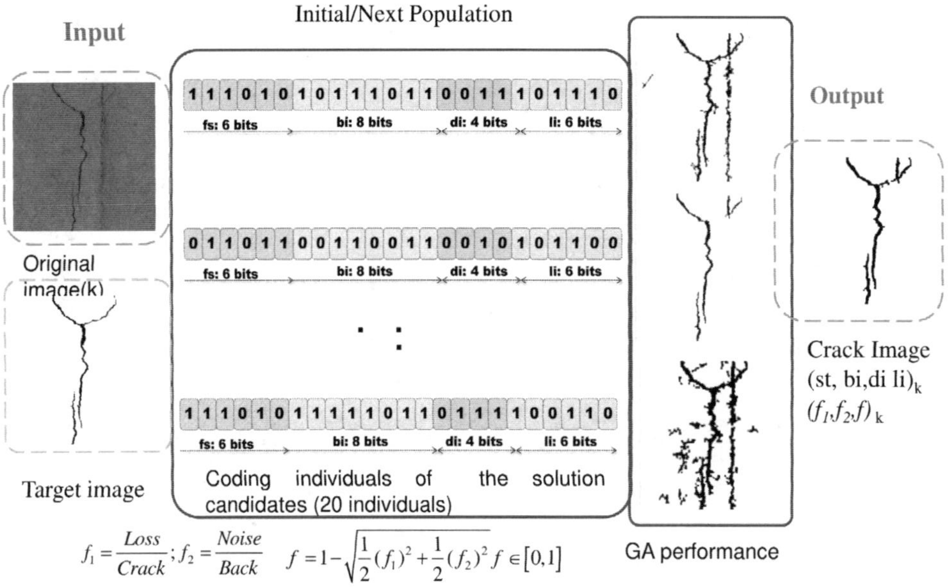

Figure 4. Procedure of genetic algorithm.

chromosome that runs until the crossover point is spliced with the part of the second parent chromosome that includes, and runs after, the crossover point shown in figure 3. The whole new generation is selected

in this manner. The mutation of bit strings ensue through bit flips at random positions. The purpose of the mutation operation is to create genotype diversification in the population in order to avoid local optimization leading to finding the best solution. Mutation point is chosen randomly. However, mutation rate is very small under 1% to avoid collapsing the genetic structure of the current population.
Step3. Stopping criterion
Evaluation of each individual meets the predefined maximum generation

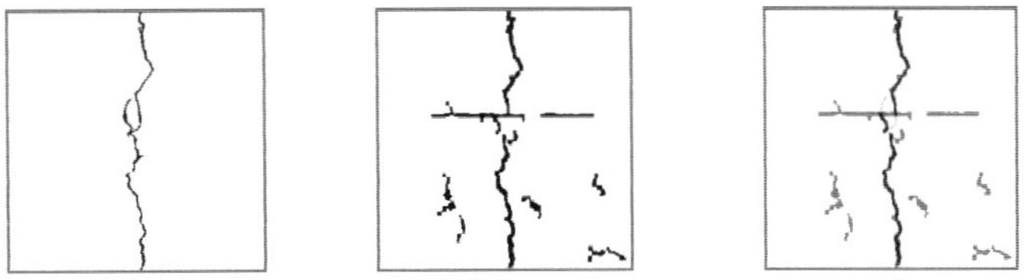

(a) (b) (c)

Figure 5. Result of supervised learning algorithm with one-pixel accuracy. (a): Ground-truth image; (b): Processed image; (c): Compared image. A number of red pixels in the compared image are noise pixels annotated Noise. A number of green pixels in the compared image are loss pixels annotated Loss.

$$f=\frac{1}{\sqrt{w_1 f_1^2 + w_2 f_2^2}},\ f \in [0\ 1]$$
$$f_1=\frac{Loss}{Crack};f_2=\frac{Noise}{Back} \tag{7}$$

Where Crack and Back are the number of black pixels and white pixels in the ground-truth image, respectively. In this paper, the weight parameters of the objective function $w_1=w_2=0.5$.
f_1, f_2 are loss rate and noise rate, respectively. f measures the accuracy of crack detection. f is larger value, the accuracy is higher.

3. Deep neural network

The output results of the image processing technique are crack images including crack pixels and crack-like non-crack pixels. The brightness of crack-like non-crack pixels is similar to the brightness of the crack pixels. The brightness is intensity value of pixel of binary image in 0 (black) or 255 (white). Figure 6 shows a crack image result of fully automated crack detection method using IPTs combined GA. As a result, the crack image contains many noise and crack like non-crack. The major challenge is how to classify them automatically in order to only keep true crack pixels.

3.1 Convolution layer

The DCNN composed of three stages. Each stage comprises of convolution layer, batch-normalization layer, max-pooling layer with drop out, and activate function (Cha et al.2017). The last layer is fully-connected layer so as to map noise candidates or crack candidates.
Table 2 indicates dimension of layers of DCNN shown in figure 1 as well as parameters of stride and padding. Where C1, C2, and C3 are the convolution layers; M-Pl, M-P2, and M-P3 are the max pooling. FCL1and FCL2 are fully-connected layers. Dr1and Dr2 are drop out with a predefined probability ratio (Tong et al.2018). In this paper, these probability ratios are 0.4.

IOP Conf. Series: Materials Science and Engineering **829** (2020) 012027 doi:10.1088/1757-899X/829/1/012027

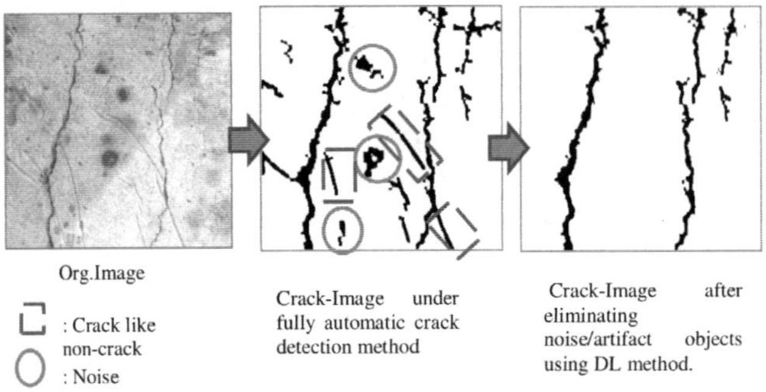

Figure 6. Application of deep learning method for true-crack pixel detection.

Table 2. Dimension of layers and parameters.

Layer Name	Number of Filter	Size	Stride	Padding	Layer Name	Number of Filter	Size	Stride	Padding
C1	32	16x16	2	2	C3	128	6x6	2	2
M-P1	1	3x3	2	-	M-P3	1	2x2	2	-
Dr1	1				Dr2	1			
C2	64	7x7	2	2	FCL1	1	128	-	-
M-P2	1	3x3	2	-	FCL2	1	2	-	-

4. Experiment results

The accuracy of the classification is expressed as the following equation:

$$ACC = \frac{TP + TN}{TP + TN + FP + FN} \tag{8}$$

Where *TP* is the total of crack images detected correctly, *TN* is the total of non-crack images detected correctly. *FP* is the total of the crack images detected incorrectly. *FN* is the total of the non-crack images detected incorrectly.

Figure 7 shows comparison of the training accuracy between the proposed method and Cha_2017 method. The result expresses that the proposed method outperforms the other, and it gains maximum accuracy at epoch 30. Further, the experiment results of the proposed training model with various image data sizes. Training maximum accuracy (*ACC*) is 96.1%, the test data accuracy is 91% with 3000 images for each class. Moreover, the minimum testing accuracy belonged to 4000 image data set is 87.5%.

5. Conclusion

This paper found the optimization of parameters of IPTs using GA. Moreover, the IPTs results combined with DCNN method resulted in high automatic crack detection. The accuracy of training model is depending on the diverse of input data. The difference of accuracy between training and testing is large. It is necessary to improve more deep CNN layers to get high score.

The final purpose was to create crack map therefore requiring the pixel-level accuracy automatically. Disadvantage of DCNN method needed a large number of input data so as to gain the high fixed accuracy. More, the computation of DCNN model is heavy relied on GPU and computer configuration.

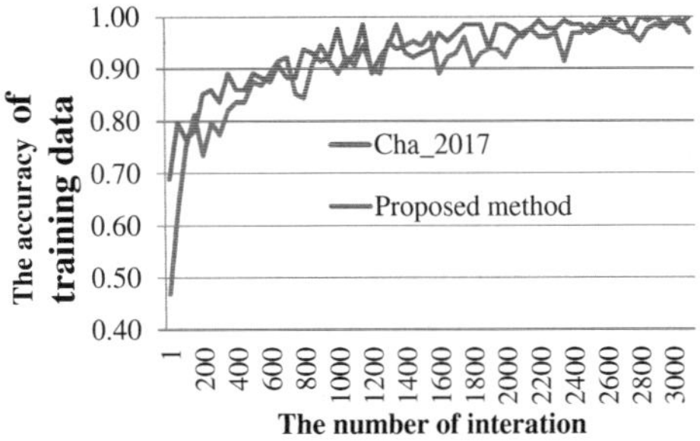

Figure 7. Comparison result between the proposed method and Cha_2017.

6. References

[1] Cha,Y., Choi,W. (2017). Deep Learning-Based Crack Damage Detection Using Convolutional Neural Networks, *Computer-aided Civil and Insfrastrure Engineering* , vol. 32, pp. 361–378.

[2] Fujita,Y., Mitani Y., and Hamamoto Y. (2006). A method for crack detection on a concrete structure, *ICPR2006: IEEE 18th Int.Conf, on Pattern Recognition*, Vol.3, New York, pp.901-904.

[3] Kawamura, K., Yoshino, K., Nakamura, H., Sato, Tarighat, A. (2013). A Valid Parameter Range Identification Method of a Digital Image Processing Algorithm for Concrete Surface Cracks Detection Using Genetic Algorithm and Decision Tree, *Proc. Japan Soc. Civ. Eng.* Vol.69.

[4] Kawamura, K., Miyamoto, Y., Nakamura, A., H., Sato, R. (2003). Proposal of a crack pattern extraction method from digital images using an interactive genetic algorithm, *Proc. Japan Soc. Civ. Eng.* Vol.72, pp.115-131.

[5] Nguyen, C., Kawamura, K., Tarighat, A. (2016) A study on semi-automatic concrete cracks detection using interactive genetic algorithm, *JCI, vol 38(1),* pp.2061-2066.

[6] Tong, Z., Gao, J., Sha, A., and Hu, L., (2018). Convolution neural network for asphalt pavement surface texture analysis, *Computer-aided civil and infrastructure engineering*, Vol 0, pp.1-17.

[7] Zhang, W., Zhang, Z., Qi, D., Liu,Y. (2014). Automatic crack detection and classification method for subway tunnel safety monitoring, *Sensors* .Vol.14, pp.19307-19328.

Acknowledgments

This research is supported by JSPS KAKENHI grant number 15K06180.

IOP Conf. Series: Materials Science and Engineering **829** (2020) 012028 doi:10.1088/1757-899X/829/1/012028

Performance of buildings around a long irregular excavation by top-down technique in downtown Nanjing

Z X Gu[1], H Li[1], Z L Meng[1], H Z Lu[1], Y Sun[1,2]

[1] College of Harbour, Coastal and Offshore Engineering, Hohai University, Nanjing, China

[2] 452860990@qq.com

Abstract. A 1257 m-long irregular deep excavation was located in the city center of Nanjing, China. The full-width and half-width top-down construction method was adopted. The deformation characteristics of the settlement of adjacent buildings based on field data is discussed, and the interrelationships between building settlement and excavation process, structural system and geological conditions are also considered. The analysis results show that building settlement is related to the excavation deformation, construction sequence, structure forms, foundation types and the distance to the foundation pit. Corner effect has played an important role in building settlement and will affect the settlement pattern eventually. The settlement surface of building outside the pit corner is three-dimensional, while the building located in the middle of the long side of the pit has a two-dimensional settlement surface.

1. Introduction

Excavation will more or less affect the surrounding buildings within the influenced zone [1]. According to previous studies, the building deformation and ground settlement outside the typical rectangular foundation pit relied on the analysis of field excavation data have been thoroughly studied [2-5]. However, few studies have been conducted on irregular foundation pits. At the same time, full-width as well as half-width construction method appear in one excavation project is rare. How to ensure the safety of construction and reduce the surrounding environment impact is a problem that needs attention. Therefore, it is necessary to monitor the buildings around the irregular foundation pit and master the law and mechanism of building deformation.

2. Project information

The project is located at Hunan Road, Nanjing, China. The soil beneath the surface is weak and has great fluidity. Figure 1 presents the layout of excavation and instrumentation. The total length of the excavation is 1257 m. Area A is a 3-story basement with excavation depth of 17.00~17.55 m, and supported by concrete struts. Area BCD is a 2-story basement with excavation depth of 13.55~15.45 m. The first layer is steel strut, and the other two layers are concrete strut. In order to ensure the traffic running, full-width construction was converted to half-width construction in part of the area CD, and the support form was adjusted to a three-layer concrete strut. The buildings surrounding the excavation include the Qingchun Theatre, the Bank of China and the Suning Global Shopping Center, etc. The surrounding buildings are mostly supported by shallow foundation. They are historic and relatively sensitive to deformation. The contents of this paper are: horizontal displacement of the retaining structure, ground settlement, and settlement of adjacent buildings.

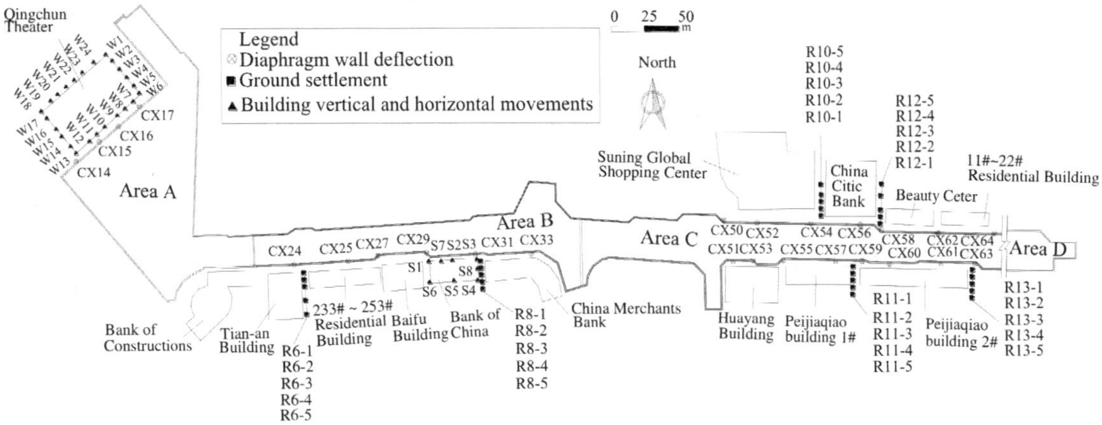

Figure 1. Layout of excavation and instrumentation.

3. Monitoring results and analysis

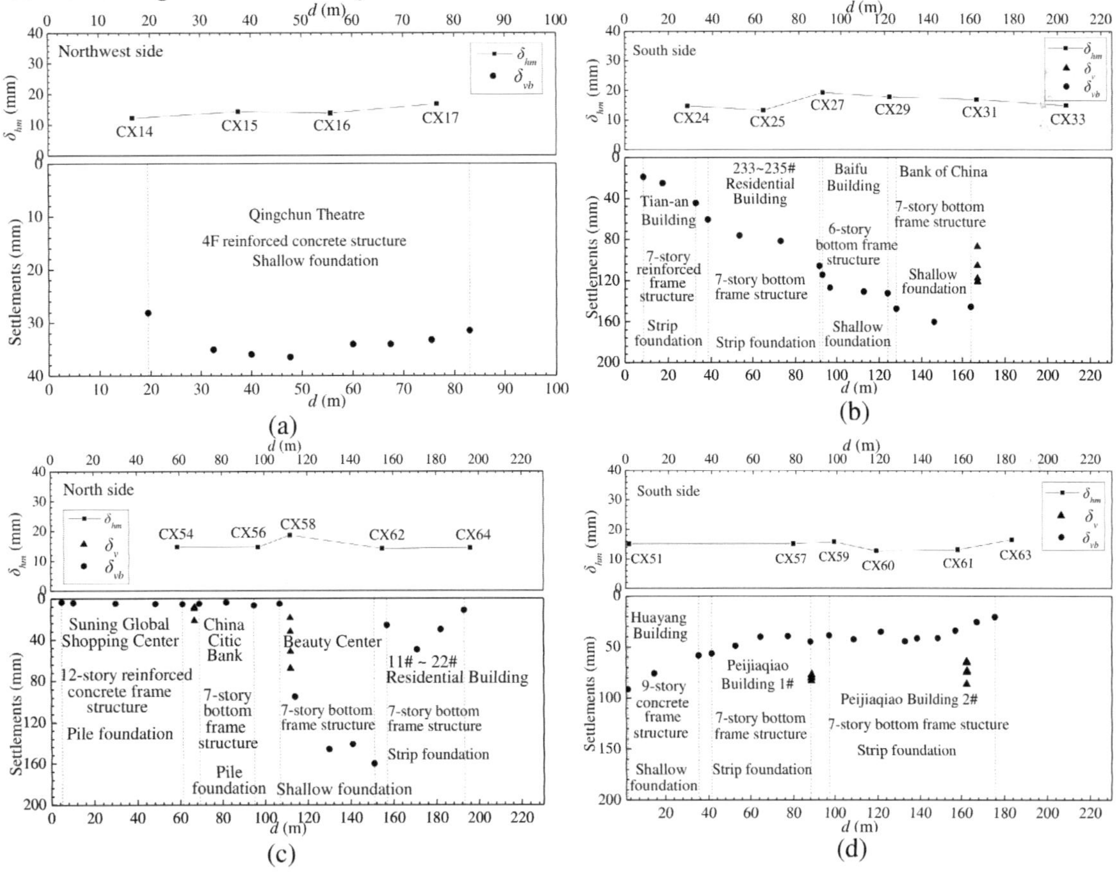

Figure 2. Realationships between δ_{hm}, δ_v and δ_{vb}: (a) North side of area A; (b) South side of area B; (c) North side of area C; (d) South side of area C.

Figure 2 reflects the maximum of the horizontal displacement of diaphragm walls (δ_{hm}), the ground settlement (δ_v) and the building settlements (δ_{vb}) along the pit. d is the distance along the alignment of the excavation. δ_{vb} had a lot to do with its foundation types and structure forms. High-rise buildings supported by shallow foundations had a large settlement. Much more δ_{vb} appeared at buildings near the convex corner than that near the concave corner. It indicates that corner effect plays a role in building settlement development. But that was not all. The significance of corner effect also depends on the length-to-depth ratio of the excavation, the depth to a relatively stiff stratum, and the stiffness of the strutting system [6]. Therefore, case analysis needs to be carried on.

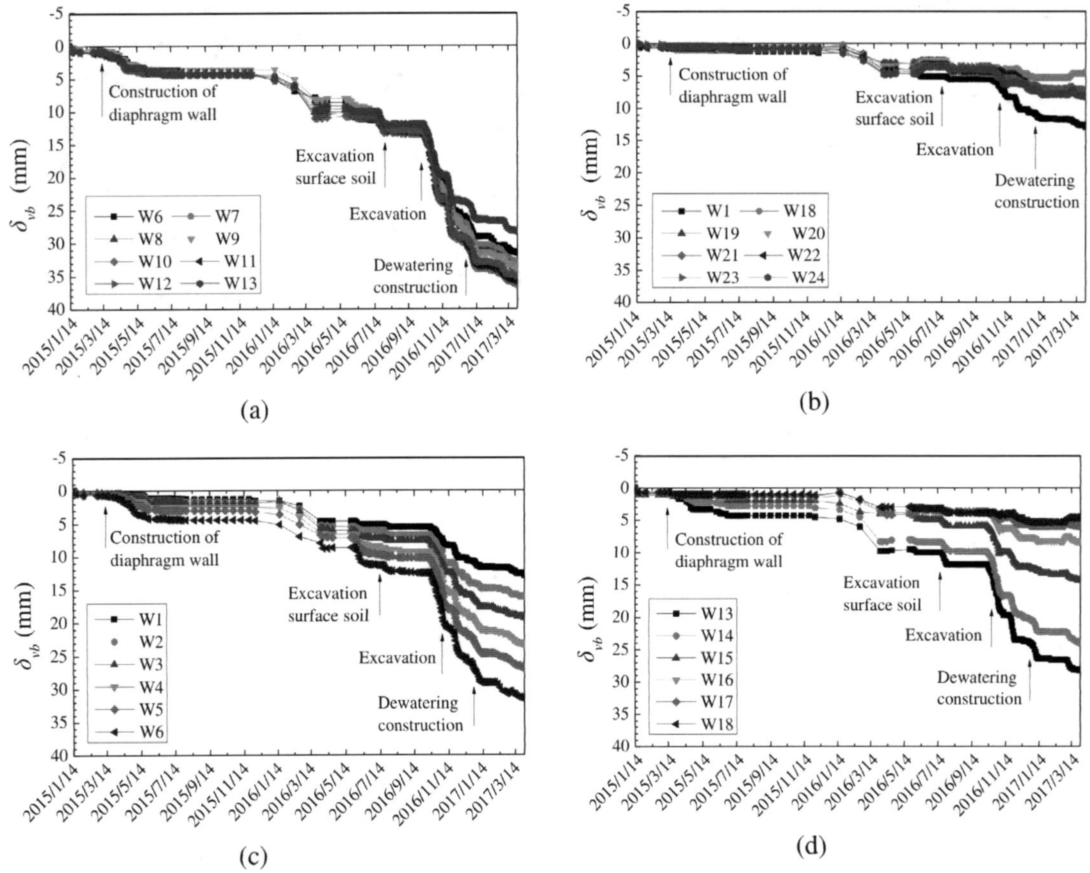

Figure 3. Settlement development of the Qingchun Theatre: (a) The east wall close to the excavation; (b) The west wall away from the excavation; (c) The north wall outside the convex corner; (d)The south wall outside the concave corner.

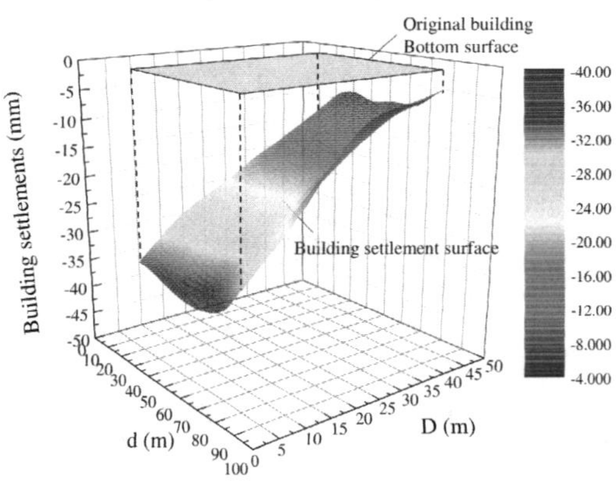

Figure 4. 3-D settlement pattern of the Qingchun Theatre.

The Qingchun Theatre is located on the west side of area A, and the outer wall of the building is about 5.4 meters away from the excavation. It is a 4-story reinforced concrete structure supported by shallow foundation. By comparing figure 3(a) and figure 3(b), the total settlement of the building wall close to the excavation was larger than that far away from the excavation. The measuring points in figure 3(c) are located outside the convex corner of the excavation, while the measuring points in figure 3(d) are located outside the concave corner of the excavation. δ_{vb} in figure 3 (d) were significantly smaller than those in figure 3(c). According to the scholars' researches [7], the convex angle effect will facilitate the soil outside the pit to settle, while the concave angle will restrict the settlement of soil. The corner effect becomes less significant as the distance to the corner. Uneven settlement occurred at both the wall parallel to the foundation pit and the wall perpendicular to the foundation pit. As a result, the settlement of the Qingchun Theatre exhibited a 3-D uneven pattern showed in figure 4, and cracks developed in several parts of the building. The joint influence of convex corner and concave corner could account for the phenomenon.

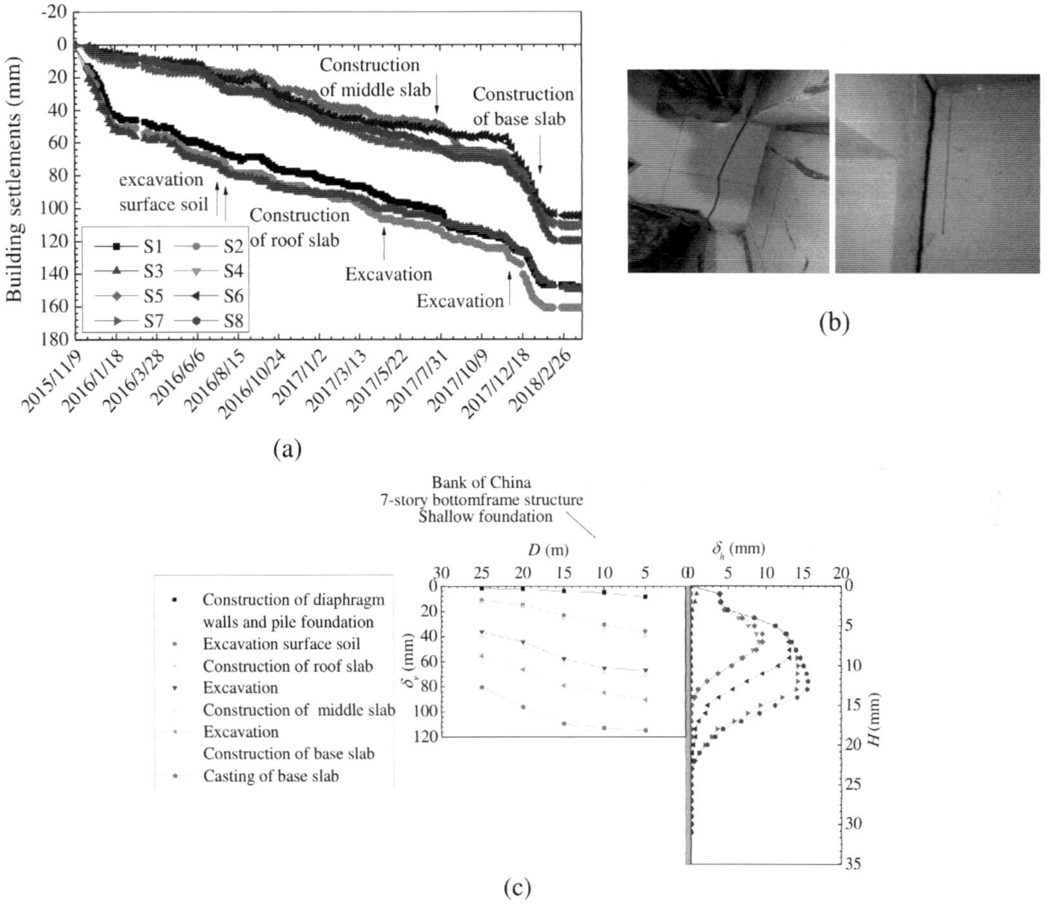

Figure 5 Bank of China: (a) Settlement developments; (b) Cracks in the second floor on the west side; and (c) Relationships between δ_{vb}, δ_v and δ_h.

The Bank of China is a 7-story bottom frame structure and supported by shallow foundation. The average distance between the building and the diaphragm wall is 5 m. During construction, the Bank of China maintained a constant settlement rate, and δ_{vb} of the north wall was larger than that of the south wall. In December 2015, due to the construction of diaphragm walls, the south wall close to the excavation settled considerably and uneven settlement increased. At this time, vertical cracks from the top to the bottom occurred on the east and west side walls. Since December 2017, the foundation pit was excavated, and the soil was weak and rheological. The whole settlement occurred uniformly around the building. Compared with the Qingchun Theatre, the Bank of China is located on the long side of the middle foundation pit, and its settlement pattern is two-dimensional.

The Bank of China was located within the influenced scope of surrounding ground settlement induced by excavation and is totally at the largest part of the ground settlement range indicated in figure 5(c). δ_h is the horizontal displacement of the diaphragm wall. During the construction of the diaphragm wall, the ground settlement and the horizontal displacement of the diaphragm wall were very small, while the building settlement on the north side was obvious. It indicates that the building is more sensitive to the construction, and the crack development as well as strength failure all promoted building settlement. As the strength of base slab increased, ground settlement and building settlement gradually tended to be stable.

The Suning Global Shopping Center is in the form of reinforced concrete frame structure. The building is 12 stories with a 2-story basement. The average distance between the building and the external diaphragm wall of excavation is 11.5 m, and the foundation form is pile foundation. The final settlement of each measuring point was very small compared to the Qingchun Theatre and the Bank of China, only in the range of 2 mm~6 mm. During the construction of the diaphragm wall, the building first settled downwards and then rose upwards. Half-width construction had little effect on the building settlement. The building is basically within the influenced scope induced by excavation. However, the maximum ground settlement was closer to the foundation pit. Thus the slope of the building would not occur. Due to the pile foundation and the reinforced concrete frame structure, the settlement of the Suning Global Shopping Center was small and few cracks occurred.

4. Conclusion

Base on the field data, most buildings are located within the influenced zone of ground settlement induced by excavation. Corner effect has a great influence on building settlement and led to a 3-D building settlement pattern. The building settlements are most obvious when the foundation pit is excavated, and tend to be stable after the construction of base slab. Both structure form and foundation form have contributed to the building settlement, and high-rise buildings supported by shallow foundations had a large settlement.

5. References

[1] Tan Y and Li M 2011 *Can. Geotech. J.* **48** 704–19
[2] Bolton M D, Lam S Y, Vardanega P J, Ng C W W, and Ma X F 2014 *Front. Struct. Civ. Eng.* **8** 201–36
[3] Kog Y C 2016 *J. Perform. Constr. Facil.* **31** 04016080
[4] Tan Y, Huang R, Kang Z, and Wei B 2016 *J. Perform. Constr. Facil.* **30** 04016040
[5] Li D, and Yan C 2018 *Adv. Civ. Eng.* 2018
[6] Zhao W, Chen C, Li S, and Pang Y 2015 *J. Intell. Robot. Syst.* **79** 351–69
[7] Pan H, Zhou C, and Cao H 2008 *Rock Soil Mech.* **29** 333–36 (In Chinese)

Acknowledgments

The research work described herein was funded by the National Natural Science Foundation of China (NSFC, Grant Number: 41672257).

AUTHOR INDEX